My Book

This book belongs to

Name: ______________________________

Cover Design by :
Gowri Vemuri

First Edition : January , 2020
Second Edition : January , 2023

Author :
Gowri Vemuri

Edited by :
Ritvik Pothapragada
Siddhartha Rangavajjula

Questions: mathknots.help@gmail.com

NOTE : ACSL is neither affiliated nor sponsors or endorses this product.

Dedication

This book is dedicated to:

My Mom, who is my best critic, guide and supporter.

To what I am today, and what I am going to become tomorrow,

is all because of your blessings, unconditional affection and support.

This book is dedicated to the

strongest women of my life ,

my dearest mom

and

to all those moms in this universe.

G.V.

American Computer Science League (ACSL) is an international computer science competition originally founded in 1978. This organization is also an institutional member of the Computer Science Teachers Association. ACSL is on the approved activities list of the National Association of Secondary School Principals (NASSP).

ACSL consists of five divisions to appeal to the varying computing abilities and interests of students. All students at a school can take the tests but can only participate in one division. A team score is the sum of the best 3 or 5 scores in each test. Those scores can come from different students with in the team, in each contest. Prizes are awarded to top scoring students and teams based on cumulative scores after the 4th test.

The **Senior / Intermediate Division** is geared to those high school students with programming experience. Each contest consists of a 30-minute, 5- questions short answer test and a take home programming problem to be solved in 72-hours. Team scores can be based on the sum of the top 3 or top 5 scores each contest.

The **Junior Division** is geared to junior high and middle school students with no previous experience programming computers. No student beyond grade 9 may compete in the Junior Division. Each contest consists of a 30-minute 5-question short answer test and a take home program to be solved in 72-hours. Team scores are based on the sum of the best 5 scores each test.

The **Classroom Division** is open to students from all grades. It consists of a selection of the non-programming problems. As its name implies, this division is particularly well-suited for use in the classroom. Each contest consists of a 50-minute, 10-question short answer test. Team scores are based on the sum of the best 5 scores each test.
The **Elementary Division** is open to students from grades 3 - 6. It consists of non-programming problems. Four categories, one each contest, will be tested. The contest consists of a 30-minute, 5-questions test.

ELEMENTARY DIVISION	CLASS ROOM / JUNIOR DIVISION	INTERMEDIATE / SENIOR DIVISI
Elementary Computer Number Systems	Computer Number Systems	Computer Number Systems
Elementary Prefix/Infix/Postfix Notation	Recursive Functions	Recursive Functions
Elementary Boolean Algebra	What Does This Program Do? Branching	What Does This Program Do?
Elementary Graph Theory	Prefix/Infix/Postfix Notation	Prefix/Infix/Postfix Notation
	Bit-String Flicking	Bit-String Flicking
	What Does This Program Do? Loops	LISP
	Boolean Algebra	Boolean Algebra
	Data Structures	Data Structures
	What Does This Program Do? Arrays	FSA/Regular Expressions
	Graph Theory	Graph Theory
	Digital Electronics	Digital Electronics
	What Does This Program Do? Strings	Assembly Language

This book is written by computer science teachers and industry experts.
Our book comprises of various practice questions in line with ACSL topics.

NOTE: ACSL is neither affiliated nor endorsed the content of this book.

This book is intended to give a hands on practice on various topics covering contest 1 and contest 2.
The questions might not replicate 100% of the actual test but are intended to reinforce the basic concepts.

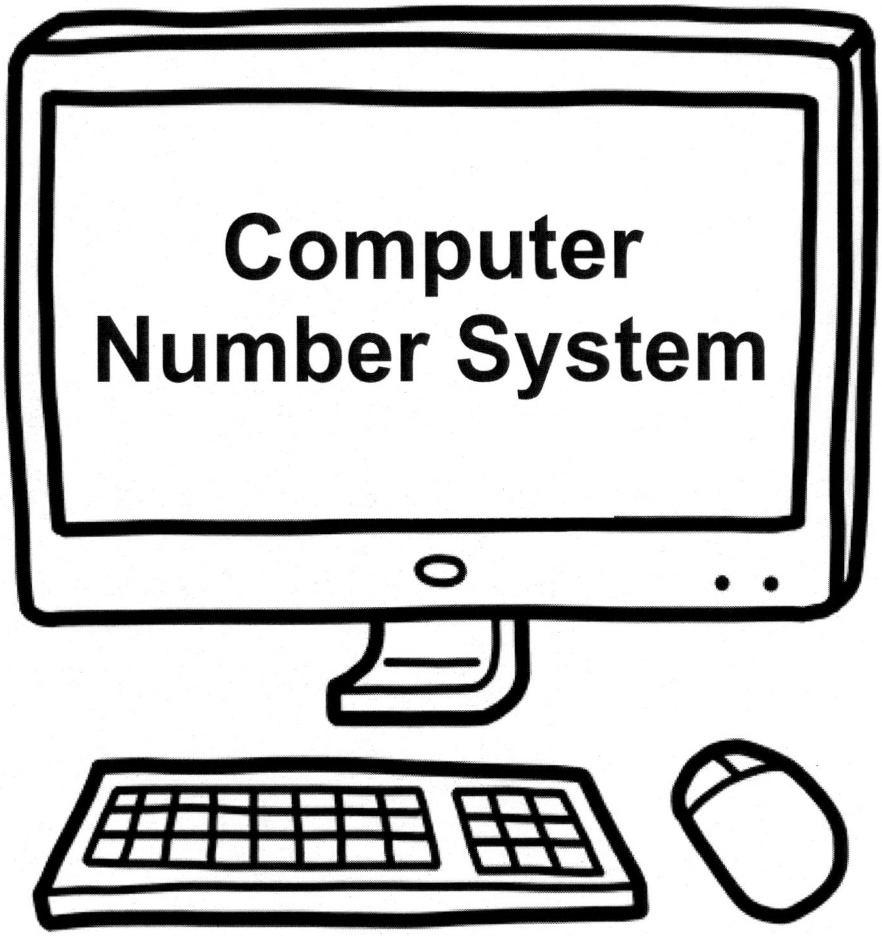
Computer
Number System

NUMBER SYSTEMS

- Decimal
- Binary
- Hexadecimal
- Octal

Decimal System:

Any number which is positionally defined in the number system is called as a decimal. All decimal numbers are with base 10.

Example :245

$= 2 \times 10^2 + 4 \times 10^1 + 5 \times 10^0 = 200 + 40 + 5$

Binary System:

As the name implies binary numbers are only two : 1 and 0 . All binary numbers are represented with base 2

Example: Convert the following to decimal system representation 0110100

64	32	16	8	4	2	1
0	1	1	0	1	0	0

$= 0X2^6 + 1X2^5 + 1X2^4 + 0X2^3 + 1X2^2 + 0X2^1 + 0X2^0$

$= 0X64 + 1X32 + 1X16 + 0X8 + 1X4 + 0X2 + 0X2$

$= 0 + 32 + 16 + 0 + 4 + 0 + 0 = 52$

Octa decimal System:

All octa decimal numbers are represented with base 8

Hexa decimal System: All octa decimal numbers are represented with base 16

Comparison chart of various Number Systems

Decimal	Binary	Octal	Hexadecimal
0	0000	0	0
1	0001	1	1
2	0010	2	2
3	0011	3	3
4	0100	4	4
5	0101	5	5
6	0110	6	6
7	0111	7	7
8	1000	10	8
9	1001	11	9
10	1010	12	A (10)
11	1011	13	B (11)
12	1100	14	C (12)
13	1101	15	D (13)
14	1110	16	E (14)
15	1111	17	F (15)

Grouping Binary Digits for Octal and Hexadecimal Numbers

EXAMPLE:

BINARY TO OCTAL

111000111_2 = 111000111 = 707_8

BINARY TO HEXA DEECIMAL

111000111_2 = 111000111 = $1C7_{16}$

COLOR REPRESENTATION :

All computer graphics are created using three main colors Red, Green and Blue. A combination of these produce various shades of the colors that we see on computer screens/graphics. Every color is represented by 6 digits with a hash sign preceding them. First two digits represent color Red, next two digits represent Green and last two digits represent Blue

White is represented by : #FFFFFF
in which first two FF is represented by Red ,
next two FF is represented by Green ,
last two FF are represented by B lue

Red Color is always represented by #FF0000
BlueColor is always represented by #00FF00
GreenColor is always represented by #0000FF

Example

FFE4B5

Red Color is represented by FF = 255

Blue Color is always represented by E4 = 228

Green Color is always represented by B5 = 181

Find the decimal values of red ,blue and green colors in $BDB76B_{16}$

BD_{16}= 11*16 + 13*1 = 189

$B7_{16}$ = 11*16 + 7*1 =183

$6B_{16}$ = 6*16 + 11*1 =107

1. Convert 148_8 to base 10.

2. Convert 109_8 to base 10.

3. Convert 287_8 to base 10.

4. Convert 427_8 to base 10.

5. Convert 538_8 to base 10.

6. Convert 649_8 to base 10.

7. Convert 1009_8 to base 10. .

8. Convert 2281_8 to base 10. .

9. Convert 10155_8 to base 10. .

10. Convert 31189_8 to base 10.   .

11. Convert 2250_8 to base 10.   .

12. Convert $2AF_{16}$ to base 10.   .

13. Convert $2AA_{16}$ to base 10. .

14. Convert $5CD_{16}$ to base 10. .

15. Convert $2DE_{16}$ to base 10.

16. Convert $A6C_{16}$ to base 10.

17. Convert $5BF_{16}$ to base 10.

18. Convert ACB_{16} to base 10. .

19. Convert $BF9_{16}$ to base 10.

20. Convert $F8D_{16}$ to base 10. .

21. Convert $FE7_{16}$ to base 10.

22. Convert the following base 8 number to base 16:

$$30521_8$$

23. Convert the following base 8 number to base 16:

$$75346_8$$

24. Convert the following base 8 number to base 16:

$$51324_8$$

25. Convert the following base 8 number to base 16:

27105_8

26. Convert the following base 8 number to base 16:

656565_8

27. Convert the following base 8 number to base 16:

135206_8

28. Convert the following base 8 number to base 16:

45623_8

29. Convert the following base 8 number to base 16:

727272_8

30. Convert the following base 8 number to base 16:

141414_8

31. Convert the following base 8 number to base 16:

$$5055055_8$$

32. Convert $C40DE_{16}$ to Octal.

33. Convert $BE6FA_{16}$ to Octal. .

34. Convert FDABC_{16} to Octal

35. Convert $\mathrm{A798C}_{16}$ to Octal

36. Convert $\mathrm{DFE5B}_{16}$ to Octal

37. Convert EEEDA_{16} to Octal

38. Convert C71F8E_{16} to Octal

39. Convert DD6BB_{16} to Octal

40. Convert $BB4FC_{16}$ to Octal

41. Convert $AEAED_{16}$ to Octal

42. How many more 1's are there in the binary representation of $9F4_{16}$ than in the binary representation of $B18_{16}$?

43. How many more 1's are there in the binary representation of $CC97_{16}$ than in the binary representation of $19AD_{16}$?

44. How many more 1's are there in the binary representation of $77F7_{16}$ than in the binary representation of $5DA2_{16}$?

45. How many more 1's are there in the binary representation of $B97F_{16}$ than in the binary representation of $3D85_{16}$?

46. How many more 1's are there in the binary representation of $55DC_{16}$ than in the binary representation of $92EA_{16}$?

47. Express the sum in hexadecimal:

$81A_{16} + 5E4_{16}$

48. Express the sum in hexadecimal:

$F9D_{16} + DEB_{16}$

49. Express the sum in hexadecimal:

$6CD_{16} + 3EE_{16}$

50. Express the sum in hexadecimal:

$EA9_{16} + DF4_{16}$

51. Express the sum in hexadecimal:

$BFC_{16} + 3DD_{16}$

52. Simplify the below in hexadecimal:

$9AD_{16} - 2CA_{16}$

53. Simplify the below in hexadecimal:

$EAD_{16} - 1FC_{16}$

54. Simplify the below in hexadecimal:

$9E7_{16} - 2BB_{16}$

55. Simplify the below in hexadecimal:

$DE3_{16} - 5CE_{16}$

56. Simplify the below in hexadecimal:

$AC9D_{16} - 1EF7_{16}$

57. Express the sum in octa decimal:

$651_8 + 339_8$

58. Express the sum in octa decimal:

$9905_8 + 5136_8$

59. Express the sum in octa decimal:

$7163_8 + 4509_8$

60. Simplify in octa decimal:

$7777_8 - 2895_8$

61. Simplify in octa decimal:

$7657_8 - 3577_8$

62. Which is larger from the below ,Order them from least to greatest

(A) 427_8 (B) $2AF_{16}$ (C) 1009_8

63. Which is larger from the below ,Order them from least to greatest

(A) $2AA_{16}$ (B) $A6C_{16}$ (C) $5BF_{16}$

64. Which is larger from the below ,Order them from least to greatest

(A) ACB   (B) 3189 C) B9F

65. Which is larger from the below , Order them from least to greatest

(A) 2281_8 (B) $2DE_{16}$ (C) $5CD_{16}$

66. Which is larger from the below, Order them from least to greatest

(A) $F8D_{16}$ (B) $FE7_{16}$ (C) 1155_8

67. What is the decimal value of the green component in the RGB code?

$DC143C_{16}$

68. Is the decimal value of the blue and green component in the RGB code given below are equal if so what are the values?

$F08080_{16}$

69. What are the decimal values of the Red, Green and Blue components in the RGB code?

$ADFF2F_{16}$

70. Is the decimal value of the blue and green component in the RGB code given below are equal ? Which one is greater and by what value?

$20B2AA_{16}$

71. Is the decimal value of the red and green component in the RGB code given below are equal ?

Which one is greater and by what value?

$DA70D6_{16}$

Prefix - Postfix

Prefix-Postfix Notations

Infix Notation:

P + Q

As a general way of writing expressions, Operators are written in-between their operands.

An expression such as A *(B + C) /D

is simplified as "First add B and C together, then multiply the result by A, then divide by D."

The Infix Notations use the general order of evaluation, using PEDMAS rules

Prefix Notation:

Infix, Postfix and Prefix notations are three different but equivalent ways of writing expressions. It is easiest to demonstrate the differences by looking at examples of operators that take two operands.

Infix notation: A + B. Prefix notation: + X Y

Operators are written before their operands.

An expression such as A *(B + C) /D = / (A *(B + C)) D

= / * A(B+C) D /= / * A + B C D

- The order of evaluation of operators is always left-to-right
  Brackets cannot be used to change this order.
- Because the "/" is to the left of the "*" in the example above, the addition must be performed before the multiplication.
- Operators act on values immediately to the left of them. For example, the "+" above uses the "B" and "C".
- Brackets can be added to make this more explicit
  (/ (* A (+ B C)) D)

SAMPLE #1:

Convert the following infix to prefix notation:

$$(\boldsymbol{A} - \boldsymbol{D}) + \frac{\boldsymbol{B}}{\boldsymbol{C}}$$

$= ((A - D)) + (\frac{B}{C})$

$= ((-AD)) + (\frac{B}{C})$

$= (-AD) + (/BC)$

$= + - AD/BC$

Postfix Notation:

Infix notation: A + B

Postfix notation (also known as "Reverse Polish notation"): A B +

Operators are written after their operands.

An expression such as A *(B + C) /D = A *(B + C) D /

= A(B+C) *D / = A B C + * D /

- The order of evaluation of operators is always left-to-right
- Brackets cannot be used to change this order.
- Because the "+" is to the left of the "*" in the example above, the addition must be performed before the multiplication.
- Operators act on values immediately to the left of them. For example, the "+" above uses the "B" and "C".
- Brackets can be added to make this more explicit

  ((A (B C +) *) D /)

Thus, the "*" uses the two values immediately preceding: "A", and the result of the addition. Similarly, the "/" uses the result of the multiplication and the "D".

In all notations, the operands occur in the same order, and just the operators have to be moved to keep the meaning correct.

SAMPLE #2:

Convert the following infix into postfix notations:

$$\frac{\boldsymbol{A+B+C\ +A^2}}{\boldsymbol{4}}$$

$$=\frac{((A+B+C)+A^2)}{4}$$

$$=\frac{((AB+C)+A^2)}{4}$$

$$=\frac{(ABC+A2\uparrow+)}{4}$$

$$=ABC+A2\uparrow+4/$$

1. Evaluate the following postfix expression. Note that all numbers are single digit.

   8 4 / 5 4 * 2 - +

2. Evaluate the following postfix expression. Note that all numbers are not single digit.

   10 / 11 + 45 * 4 - +

3. Evaluate the following postfix expression. Note that all numbers are not single digit.

   10 5 / 9 3 * 3 − +

4. Evaluate the following postfix expression. Note that all numbers are not single digit.

   44 4 / 10 2 * 7 − +

5. Evaluate the following postfix expression. Note that all numbers are not single digit.

   36 6 / 4 4 * 9 − +

6. Evaluate the following postfix expression. Note that all numbers are not one digit.

   7 7 * 3 4 + / 14 2 * 3 -

7. Evaluate the following postfix expression if all numbers are single digits.

5 2 4 + 2 ^ * 8 4 / 2 * −

8. Evaluate the following postfix expression if all numbers are single digits.

9 1 1 + 4 ^ * 12 6 / 7 * −

9. Evaluate the following postfix expression if all numbers are single digits.

1 1 2 + 2 ^ * 2 2 / 1 * −

10. Evaluate the following postfix expression if all numbers are not single digits.

5 1 5 + 2 ^ * 15 5 / 6 * −

11. Evaluate the following postfix expression if all numbers are not single digits.

    7 4 3 - 9 ^ * 39 3 / 2 * +

12. Evaluate the following prefix expression. Note that all numbers are single digits in the expression.

    + / + 6 6 6 ^ 4 1

13. Evaluate the following prefix expression if all numbers are single digits.

- * + 4 6 8 * 4 / 8 4

14. Evaluate the following prefix expression. Note that all numbers are single digits in the expression.

- * + 1 2 4 ^ 4 1

15. Evaluate the following prefix expression if all numbers are single digits.

− * + 3 1 2 * 4 / 6 3

16. Evaluate the following prefix expression. Note that all numbers are single digits in the expression.

- * + 3 2 5 ^ 3 1

17. Evaluate the following prefix expression if all numbers are single digits.

- * + 7 1 5 * 3 / 9 3

18. Evaluate the following prefix expression. Note that all numbers are single digits in the expression.

* ^ / 6 3 2 + 8 2

19. Evaluate the following prefix expression if all numbers are single digits.

    - ^ + 5 5 2 * 5 / 4 2

20. Evaluate the following prefix expression. Note that all numbers are single digits in the expression.

    + / + 3 3 3 ^ 3 1

21. Evaluate the following prefix expression if all numbers are single digits.

− * + 9 3 6 * 4 / 6 2

22. Evaluate the following prefix expression if all numbers are not single digits.

+ * / + 20 5 5 ^ 2 / 12 4 35

23. Convert the infix expression to a postfix expression. Perform all operations in the same order as the given expression.

8 * 3 / 6 − 4 ^ 1 - 2 * 3 + 3 ^ 3

24. Convert the infix expression to a postfix expression. Perform all operations in the same order as the given expression.

(8 + 1) / 9 * 8 / (8 − 2)

25. Convert the infix expression to a postfix expression. Perform all operations in the same order as the given expression.

9 − 2 * 7 + 3 * 1 + 5

26. Convert the infix expression to a postfix expression. Perform all operations in the same order as the given expression.

(9 - 1) ^ 4 * (7 + 3) ^ 4 / 5

27. Convert the infix expression to a prefix expression. Perform all operations in the same order as the given expression.

32 – 9 * 7 + 8 * 5 – 1

28. Convert the infix expression to prefix. Perform all operations in the same order as the expression.

7 / (5 – 4) + (8 + 2)^5 + 8 / 4

29. Convert the infix expression to a prefix expression. Perform all operations in the same order as the given expression.

$10^{(4 - 2)} \quad * \quad (7 + 3)^4$

30. Convert the infix expression to a prefix expression. Perform all operations in the same order as the given expression.

$8^{(8 - 5)} \quad / \quad (6 + 4)^7$

31. Translate the following postfix expression to prefix if all numbers are single digits.

6 4 + 3 * 6 3 4 + * −

32. Translate the following postfix expression to prefix if all numbers are single digits.

4 1 − 7 4 + 1 * 7 1 4 + * −

33. Translate the following postfix expression to prefix if all numbers are single digits.

3 2 + 1 * 3 2 1 + / -

34. Translate the following postfix expression to prefix if all numbers are single digits.

1 1 + 2 * 1 1 2 + * -

35. Translate the following prefix expression to postfix. Note that all numbers are single digit s in the expression.

* ^ + 3 2 3 + 2 3

36. Translate the following prefix expression to postfix. Note that all numbers are single digits in the expression.

/ ^ + 8 4 8 – 4 1

37. Translate the follow postfix expression to prefix. Note that all numbers are single

7 6 − 6 1 + + 4 5 * 3 − +

38. Translate the follow postfix expression to prefix. Note that all numbers are single digits.

4 3 * 2 4 + + 2 4 * 3 − +

39. Translate the following prefix expression to postfix. Note that all numbers are single digits.

− / + 2 2 1 * * 2 1 2

40. Translate the following prefix expression to postfix. Note that all numbers are single digits.

+ / + 6 7 6 * * 7 6 7

41. Translate the following prefix expression to postfix. Note that all numbers are single digits.

+ / - 8 7 5 * * 5 6 4

42. Translate the following prefix expression to postfix. Note that all numbers are single digits.

+ / - 9 3 3 * - 9 3 9

43. Translate the following prefix expression to postfix. Note that all numbers are single digits.

+ / + 4 5 6 * − 5 6 4

44. Translate the following prefix expression to postfix. Note that all numbers are single digits.

* / − 9 1 2 * + 9 5 * 8
3

45. Translate the following prefix expression to postfix. Note that all numbers are single digits

* / – 7 5 4 * – 7 1 + 6 3

Boolean Algebra

Boolean Algebra

Boolean Algebra is the mathematics we use to analyze digital gates and circuits. We can use these "Laws of Boolean" to both reduce and simplify a complex Boolean expression in an attempt to reduce the number of logic gates required.
Boolean Algebra is a system of mathematics based on logic that has its own set of rules or laws which are used to define and reduce Boolean expressions.
The variables used in **Boolean Algebra** only have one of two possible values, 0" and "1" but an expression can have an infinite number of variables.
Each variable can have a value of 1 or 0 only.

Laws of Boolean Algebra :

1. Annulment Law – A term AND´ed with a "0" equals 0 or OR´ed with a "1" will equal 1

→ A . 0 = 0 A variable AND'ed with 0 is always equal to 0
→ A + 1 = 1 A variable OR'ed with 1 is always equal to 1

2. Identity Law – A term OR´ed with a "0" or AND´ed with a "1" will always equal that term

→ $A + A = A$
→ $AA = A$
→ A + 0 = A A variable OR'ed with 0 is always equal to the variable
→ A . 1 = A A variable AND'ed with 1 is always equal to the variable

3. Idempotent Law – An input that is AND´ed or OR´ed with itself is equal to that input

→ A + A = A A variable OR'ed with itself is always equal to the variable
→ A . A = A A variable AND'ed with itself is always equal to the variable

4. <u>Complement Law</u> – A term AND´ed with its complement equals “0” and a term OR´ed with its complement equals “1”

→ $\overline{A} . A = 0$ A variable AND'ed with its complement is always equal to 0
→ $\overline{A} + A = 1$ A variable OR'ed with its complement is always equal to 1

5. <u>Commutative Law</u> – The order of application of two separate terms is not important

→ $A . B = B . A$ The order in which two variables are AND'ed makes no difference

→ $A + B = B + A$ The order in which two variables are OR'ed makes no difference

6. <u>Double Negation Law</u> – A term that is inverted twice is equal to the original term

→ $A = A$ A double complement of a variable is always equal to the variable

7. <u>de Morgan´s Theorem</u> – There are two “de Morgan´s” rules or theorems,

→ Two separate terms NOR´ed together is the same as the two terms inverted (Complement) and AND´ed for example: $A+B = A . B$

8. **De Morgan's Theorem**

$\overline{(A + B)} = \overline{A}\,\overline{B}$

→ Two separate terms NAND´ed together is the same as the two terms inverted (Complement) and OR´ed for example: A.B = A + B

$\overline{A}\,\overline{B} = \overline{(A + B)}$

Other algebraic Laws of Boolean not detailed above include:

9. <u>Distributive Law</u> – This law permits the multiplying or factoring out of an expression.

→ A(B + C) = A.B + A.C (OR Distributive Law)
→ A + (B.C) = (A + B).(A + C) (AND Distributive Law)

10. <u>Absorptive Law</u> – This law enables a reduction in a complicated expression to a simpler one by absorbing like terms.

→ A + (A.B) = A (OR Absorption Law)
→ A(A + B) = A (AND Absorption Law)

11. <u>Associative Law</u> – This law allows the removal of brackets from an expression and regrouping of the variables.

→ A + (B + C) = (A + B) + C = A + B + C (OR Associate Law)
→ A(B.C) = (A.B)C = A . B . C (AND Associate Law)

12. $AB + A\overline{B} = A$
$(A + B)(A + \overline{B}) = A$

13. $A + AB = A$
$A(A + B) = A$

14. $A + \overline{A}B = (A + B)$

$A(\overline{A} + B) = AB$

15. $A \oplus B = A\overline{B} + \overline{A}B$
The *xor* of two values is true whenever the values are different. It uses the $\oplus$ operator, and can be built from the basic operators:

$A \oplus B = A\overline{B} + \overline{A}B$
The values of *xor* for all possible inputs are shown in the truth table below

A	B	$A \oplus B$
0	0	0
0	1	1
1	0	1
1	1	0

16. The *xnor* of two values is true whenever the values are the same. It is the *not* of the *xor* function. It uses the $\odot$ operator: $A \odot B = \overline{A \oplus B}$
The *xnor* can be built from basic operators:
$A \odot B = \overline{A \oplus B} = AB + \bar{A}\bar{B}$
The values of *xnor* for all possible inputs is shown in the truth table below:

A	B	$A \odot B$
0	0	1
0	1	0
1	0	0
1	1	1

Example :
Simplify the following expression: (A + B)(A + C)

(A + B)(A + C) = A.A + A.C + A.B + B.C (Distributive law)

= A + A.C + A.B + B.C (Idempotent AND Law)

= A(1 + C) + A.B + B.C (Distributive law)

= A.1 + A.B + B.C (Identity OR law)

= A(1 + B) + B.C (Distributive law)

= A + B.C (Identity AND law)

(A + B)(A + C) = A + B.C

Evaluate the following expression as either TRUE or FALSE:

1. (((12 + 7) <= 18) AND ((13 + 3)>= 4*3)) OR NOT (9 * 1 <= 5 * 0)

2. ((8/2) >= 5) AND (2 * 0) >= 1 * 1) OR NOT (9/3 > 2 * 2)

3. ((8 - 8) <= (8 + 8) OR ((16/4) >= 3 * 4) OR NOT (6 * 2 >= 2 * 6)

4. ((2* (7 + 3) <= 3 * (1 + 7) AND ((5 + 5) >= 4 * 1) OR NOT (63/7 > = 3 * 3)

5. ((3/8 <= 2/7) AND ((1/6) >= 5 * (1/10)) OR NOT (21 - 8 = 2 + 9)

6. NOT (4^2 < 2 * 4) OR (45/ 9 ≥ 4/2 AND 55 * 2 < 10 * 5)

7. NOT (8^2 < 2^3) OR (100 / 25 ≥ 0 AND 108 * 0 > 100 * 8)

8. NOT (54/2 -1 < = 26) OR (99-90 ≥ 100-99 AND 9^2 < 9 *9)

9. NOT (11^2 < 2 * 11) OR (55-50 ≥ 3 * 7 AND 130/ 2 > 13 * 5)

Evaluate the following expression as either TRUE or FALSE:

10. NOT (2 * (30 + 25) < = 55 + 55) OR (((69 + 23) + 10 = 69 + (23 + 10)) AND ((14*12) * 16 <= 14* (12 *16))

Write the Truth table for the below

11. How many ordered pairs make the following statement FALSE?

(A OR B) OR (NOT (A OR B) OR NOT (A AND B))

Write the Truth table for the below

12. NOT A OR (A OR (A OR B))

Evaluate the following expression as either TRUE or FALSE:

13. NOT A OR (A OR (A AND B))

Write the Truth table for the below

14. A OR NOT (A OR B)

Write the Truth table for the below

15. A AND NOT (A AND B)

Evaluate the following expression as either TRUE or FALSE:

16. NOT A AND (A AND (A AND B))

Write the Truth table for the below

17. (A OR NOT NOT B) OR (A OR NOT NOT A) AND (NOT (A AND B))

Write the Truth table for the below

18. ((B AND NOT NOT B) AND (A OR NOT NOT A)) AND (A AND NOT B)

Write the Truth table for the below

19. (A OR B) AND (B AND A) OR (A AND B)

Write the Truth table for the below

20. ((A AND B) AND (B AND A)) OR (A OR B)

Write the Truth table for the below

21. ((A AND B) OR (B AND NOT B)) AND (B AND NOT NOT B)

Write the Truth table for the below

22. ((A AND B) AND (B AND NOT B)) OR (A AND NOT NOT A)

Write the Truth table for the below

23. ((A AND B) AND (B AND NOT B)) AND (B AND NOT NOT B)

Write the Truth table for the below

24. ((A OR B) AND (B OR NOT B)) AND (B OR NOT NOT B)

Write the Truth table for the below

25. (NOT ((A AND A) AND (B OR B))) AND NOT (A AND B)

Write the Truth table for the below

26. (NOT ((A OR A) OR (B OR B))) OR NOT (A OR B)

Write the Truth table for the below

27. (NOT ((A OR A) AND (B OR B))) AND NOT (A OR B)

Evaluate the following expression as either TRUE or FALSE:

28. ((NOT A OR A) AND (B OR NOT B)) OR NOT (A OR B)

Write the Truth table for the below

29. ((A AND NOT NOT A) OR (A OR NOT NOT A)) AND NOT (A AND B)

Write the Truth table for the below

30. ((A OR NOT A) AND (B OR NOT B)) AND (B OR NOT A)

Write the Truth table for the below

31. ((B OR NOT A) AND (A OR NOT A)) AND (A OR NOT B)

Write the Truth table for the below

Simplify the following Boolean expression:

32. (~A + ~B) (A + B)

33. A (~A + B) (~A + ~B)

34. ~ (~A + B) (A + ~B)

35. Is the below statement TRUE or FALSE?

~A . ~B + B (A + ~B)

36. ~ [A (B + ~B) (A + ~A) + B]

37. ~[A (B + ~B) (A + ~A)] + ~B

38. ~[A (B + ~B) (A + ~A)] + B

39. A.(~A + B) + A ~B + ~ A

40. B[~A(~A + ~B) + A .~B] + A] + ~B

41. [A (B + ~B) (A + ~A)] + (~~A + ~A)(~A)

42. [~A (B + ~B) (A + ~A)] + (~~A + ~A)(A)

43. A[~A (B + ~B) (A + ~A)] + (~~A)(~A)

44. ~B (~AB + B) + ~A~B

45. B (~A~B + ~B) + ~B (B + ~B) + B

46. $(A + \sim B)(\sim A + B) + A(\sim B) + B\sim A$

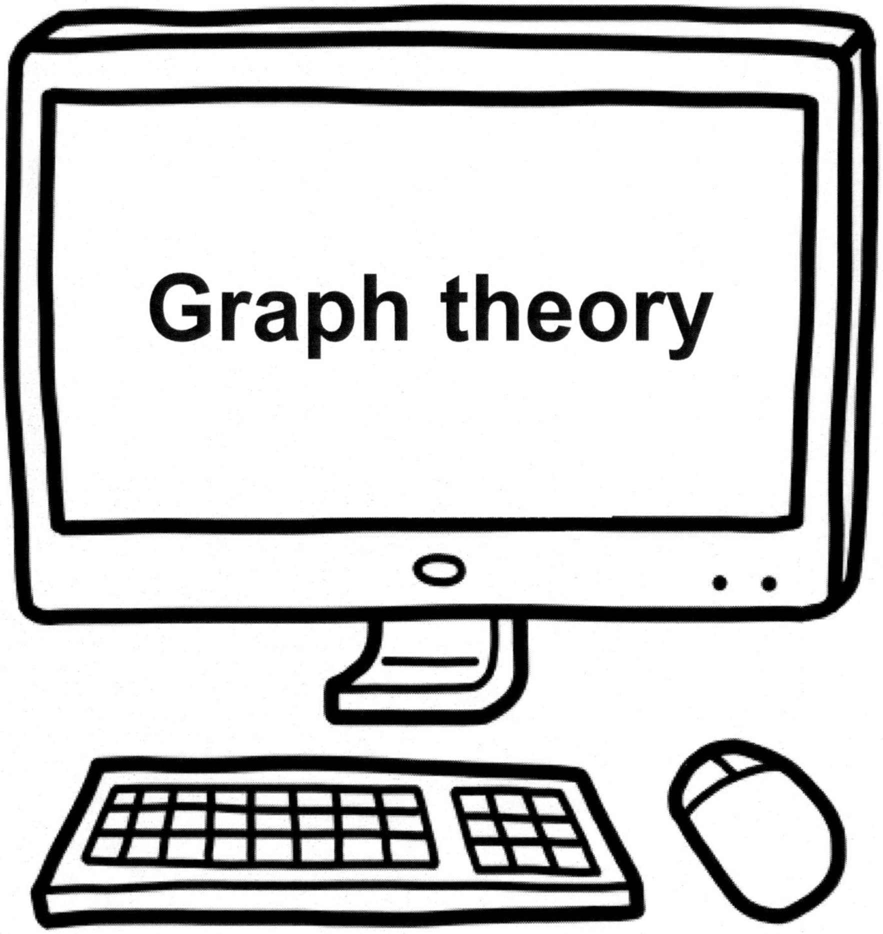
Graph theory

GRAPH THEORY:

Graph theory is used in many situations in a day to day life.

Example:
#1 Minimum cost of a underground cable wire to various locations from a central location.
#2 Fastest route from national capital to each state of the capital.
#3 Asports league scheduled into minimum number of weeks

A graph has vertices and edges. A loop is an edge whose endpoints are equal.
A simple graph is a graph having no loops or multiple edges.

DATA Structures:

Array form
root node i = 0
left child : 2i + 1
right child: 2i + 2

https://video.search.yahoo.com/yhs/search?fr=yhs-sz-002&hsimp=yhs-002&hspart=sz&p=tree+data+structure#id=8&vid=e915476b849f7eac1ca0afa34a1f8571&action=view

1) Green island has 12 visiting places connected by bridges to walk as given below. Jack can go from A to B , B to C , D to C , D to E , F to E , F to G , G to H , J to I , I to H , K to J , A to L , and L to K. Draw a graph showing the map with connections to all the bridges in Green island.

2) Green island has 12 visiting places connected by bridges to walk as given below. Jack can go from A to B , B to C , D to C , D to E , F to E , F to G , G to H , J to I , I to H , K to J , A to L , L to K , and F to L. Draw a graph showing the map with connections to all the bridges in Green island.

3) Green island has 12 visiting places connected by bridges to walk as given below. Jack can go from A to B , B to C , D to C , D to E , F to E , F to G , G to H , J to I , I to H , K to J , A to L , L to K , C to I and F to L. Draw a graph showing the map with connections to all the bridges in Green island.

4) Green island has 12 visiting places connected by bridges to walk as given below. Jack can go from A to B , B to C , D to C , D to E , F to E , F to G , G to H , J to I , I to H , K to J , A to L , L to K , C to I , A to K and F to L. Draw a graph showing the map with connections to all the bridges in Green island.

5) Crazy park has 12 rides A , B , C , D , E , F , G , H , I , J , K , L.
The rides are connected to each other by a certain walking path as below.
A to B , B to C , D to C , D to E , F to E , F to G , G to H , J to I , I to H , K to J , A to L , L to K , C to I , A to K , F to L , and D to B
Draw a graph showing the map with the path to all the rides.

6) Crazy park has 12 rides A , B , C , D , E , F , G , H , I , J , K , L.
The rides are connected to each other by a certain walking path as below.
A to B , B to C , D to C , D to E , F to E , F to G , G to H , J to I , I to H , K to J , A to L , L to K , C to I , A to K , F to L , D to B , J to H , and G to E .
Draw a graph showing the map with the path to all the rides.

7) Crazy park has 12 rides A , B , C , D , E , F , G , H , I , J , K , L.
The rides are connected to each other by a certain walking path as below.
A to B , B to C , D to C , D to E , F to E , F to G , G to H , J to I , I to H , K to J , A to L , L to K , C to I , A to K , D to B , J to H , and G to E .
Draw a graph showing the map with the path to all the rides.

8) Crazy park has 12 rides A , B , C , D , E , F , G , H , I , J , K , L.
The rides are connected to each other by a certain walking path as below.
A to B , B to C , D to C , D to E , F to E , F to G , G to H , J to I , I to H , K to J , A to L , L to K , A to K , D to B , J to H , and G to E .
Draw a graph showing the map with the path to all the rides.

9) Julia has 4 horse stables in her farm. Each stable is connected to the other by a road as given below
A to B , B to C , D to C , D to E , F to E , F to G , G to H , J to I , I to H , K to J , A to L , L to K , A to K , D to B , I to L , J to H , and G to E.
Draw a graph showing the map with connecting the horse stables at Julia's farm.

10) Julia has 4 horse stables in her farm. Each stable is connected to the other by a road as given below
A to B , B to C , D to C , D to E , F to E , F to G , G to H , J to I , I to H , K to J , A to L , L to K , A to K , D to B , I to L , C to F , J to H , and G to E.
Draw a graph showing the map with connecting the horse stables at Julia's farm.

11) Julia has 4 horse stables in her farm. Each stable is connected to the other by a road as given below
A to B , B to C , D to C , D to E , F to E , F to G , G to H , J to I , I to H , K to J , A to L , L to K , A to K , D to B , I to L , L to C , C to F , J to H , and G to E. Draw a graph showing the map with connecting the horse stables at Julia's farm.

12) Julia has 4 horse stables in her farm. Each stable is connected to the other by a road as given below
A to B , B to C , D to C , D to E , F to E , F to G , G to H , J to I , I to H , K to J , A to L , L to K , A to K , D to B , I to L , L to C , C to F , I to F , J to H , and G to E. Draw a graph showing the map with connecting the horse stables at Julia's farm.

13) Draw an undirected graph with the vertices A,B,C,D,E,F,G,H,I,J,K , and L. Edges as AB , BC , CD , BD , CF , FE , ED , EG , GF , FI , HG , HI , HJ , IJ , JK , IL , KL , AL , and LC

14) Draw an undirected graph with the vertices A,B,C,D,E,F,G,H,I,J,K , and L. Edges as AB , BC , CD , CF , FE , ED , EG , GF , FI , HG , HI , HJ , IJ , JK , IL , KL , AL , and LC

15) Draw an undirected graph with the vertices A,B,C,D,E,F,G,H,I,J,K , and L. Edges as AB , BC , CD , CF , FE , ED , EG , GF , FI , HG , HI , IJ , JK , IL , KL , AL , and LC

16) Draw an undirected graph with the vertices A,B,C,D,E,F,G,H,I,J,K , and L. Edges as AB , BC , CD , CF , FE , ED , GF , FI , HG , HI , IJ , JK , IL , KL , AL , and LC

17) Draw an undirected graph with the vertices A,B,C,D,E,F,G,H,I,J,K , and L.
Edges as AB , BC , CD , CF , FE , ED , GF , FI , HG , HI , AF
IJ , JK , IL , KL , AL , and LC

18) Draw an undirected graph with the vertices A,B,C,D,E,F,G,H,I,J,K , and L.
Edges as AB , BC , CD , CF , FE , ED , GF , FI , HG , HI , AF , BI
IJ , JK , IL , KL , AL , and LC

19) Draw an undirected graph with the vertices A,B,C,D,E,F,G,H,I,J,K , and L.
Edges as AB , BC , CD , CF , FE , ED , GF , FI , HG , HI , AF , BI , GL
IJ , JK , IL , KL , AL , and LC

20) Draw an undirected graph with the vertices A,B,C,D,E,F,G,H,I,J,K , and L.
Edges as AB , BC , CD , CF , FE , ED , GF , FI , HG , HI , AF , BI , GL , CH
IJ , JK , IL , KL , AL , and LC

21) How many paths of length equals to 2 exist in the below figure ? List all of them.

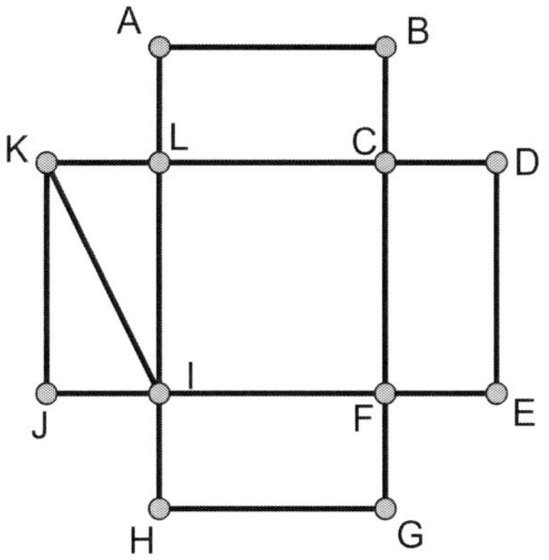

22) How many paths of length equals to 2 exist in the below figure ? List all of them.

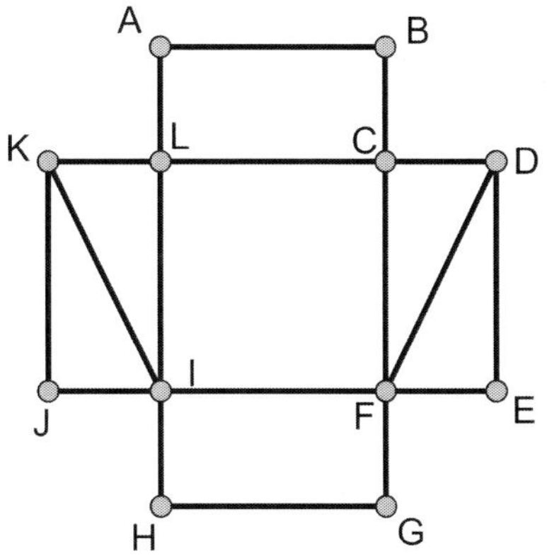

23) How many paths of length equals to 2 exist in the below figure ?
List all of them.

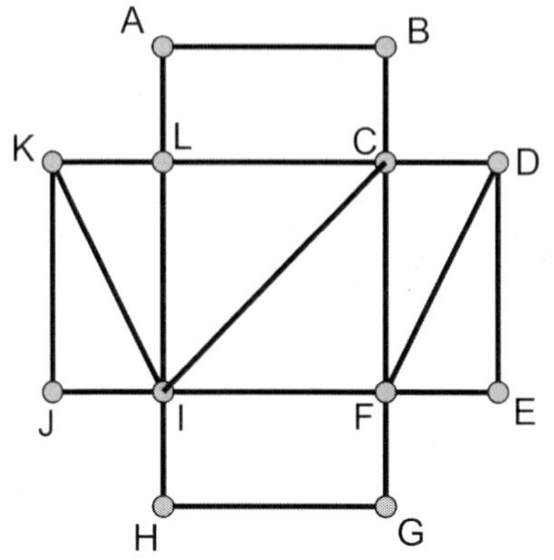

24) How many paths of length equals to 2 exist in the below figure ?
List all of them.

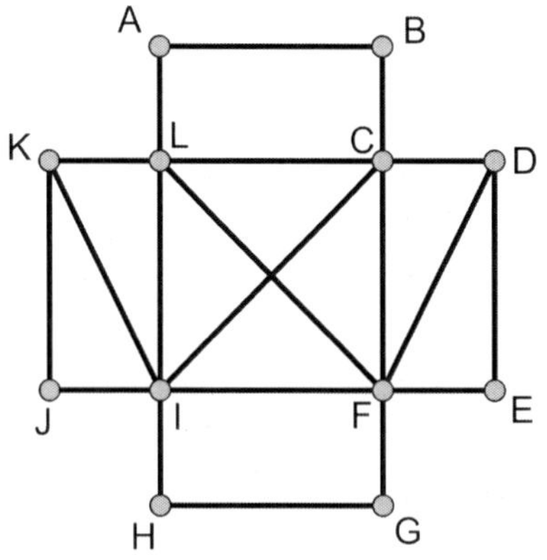

25) How many paths of length equals to 3 originate from the vertex O in the below figure ? List all of them.

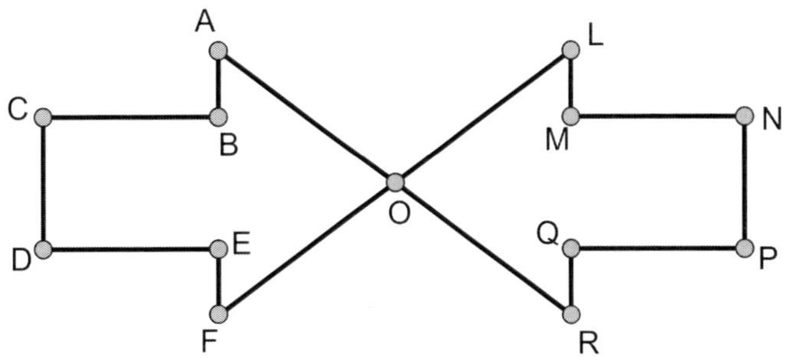

26) How many paths of length equals to 3 originate from the vertex O in the below figure ? List all of them.

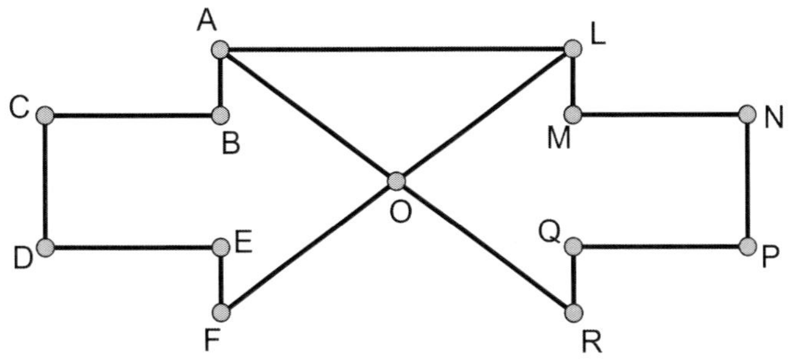

27) How many paths of length equals to 3 originate from the vertex O in the below figure ? List all of them.

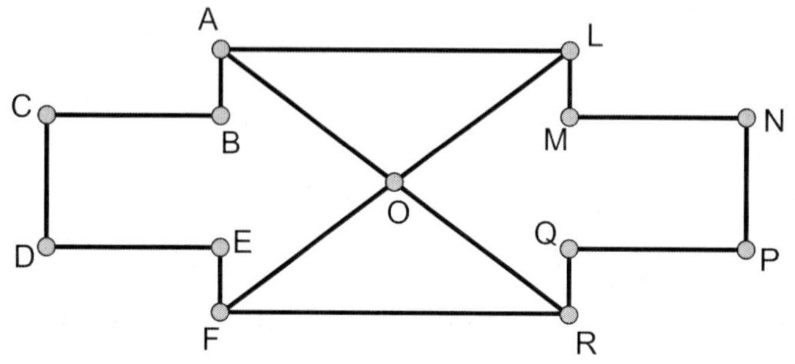

28) How many paths of length equals to 2 in the below figure ? List all of them.

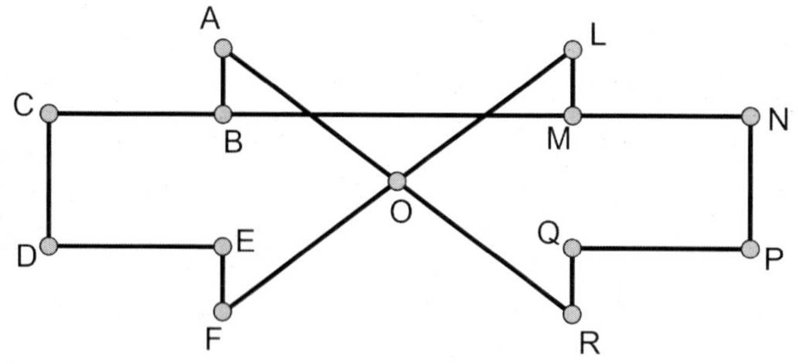

29) How many paths of length equals to 2 in the below figure ? List all of them.

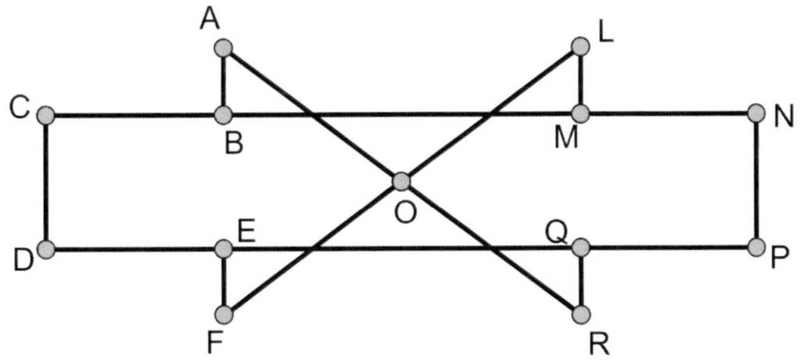

30) How many paths of length equals to 2 in the below figure ? List all of them.

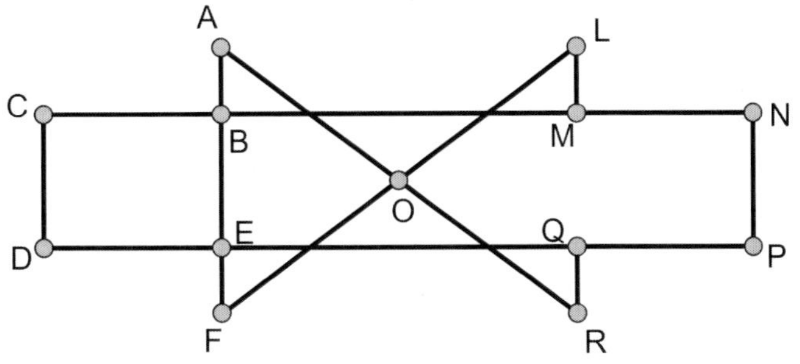

31) How many paths of length equals to 2 in the below figure ? List all of them.

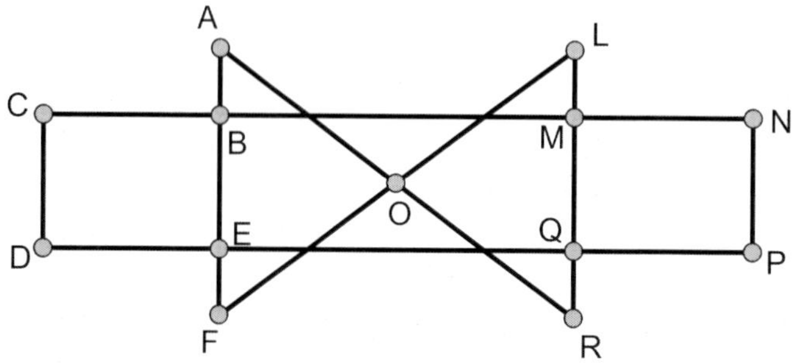

32) Find if the following graph is traversable or not ?

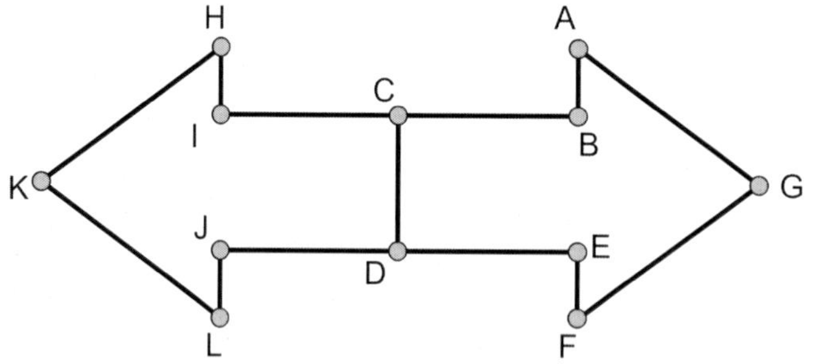

33) Find if the following graph is traversable or not ?

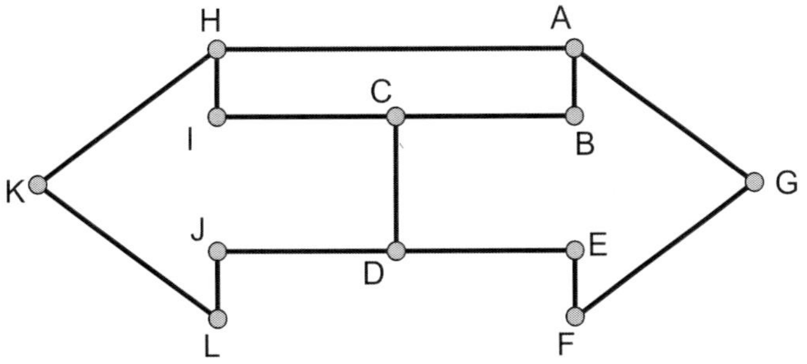

34) Find if the following graph is traversable or not ?

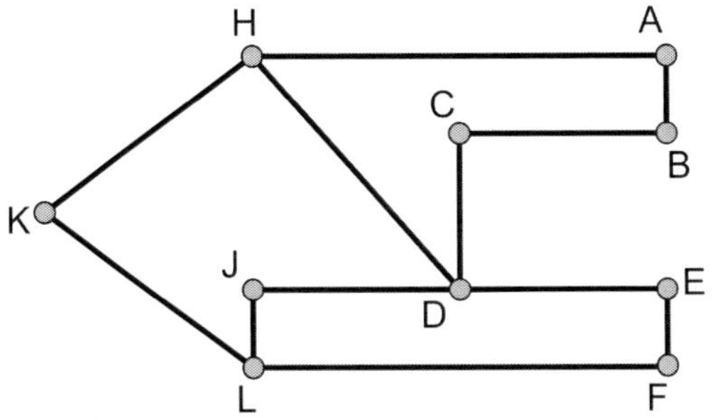

35) Find if the following graph is traversable or not ?

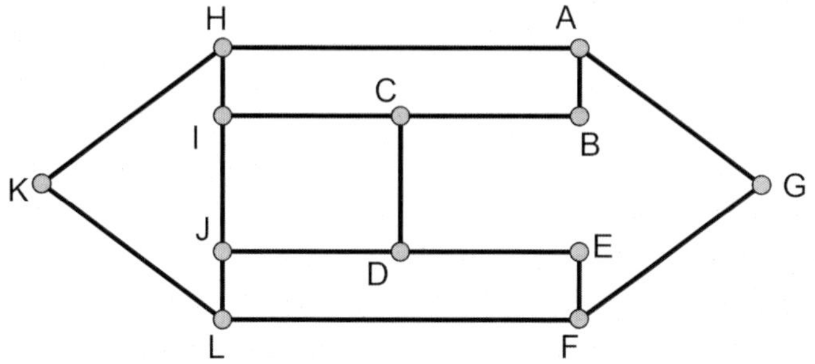

36) Find if the following graph is traversable or not ?

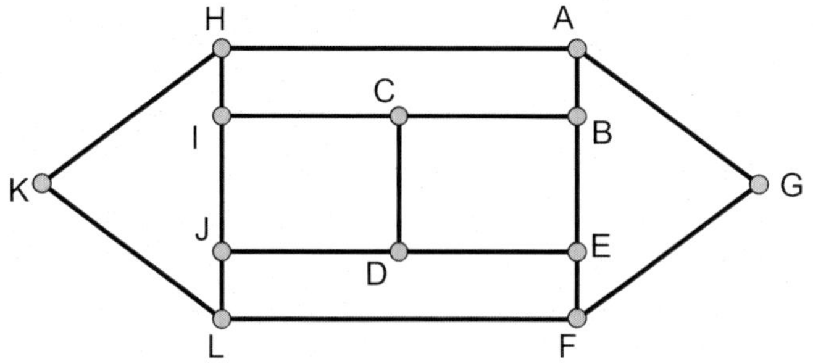

37) Find if the following graph is traversable or not ?

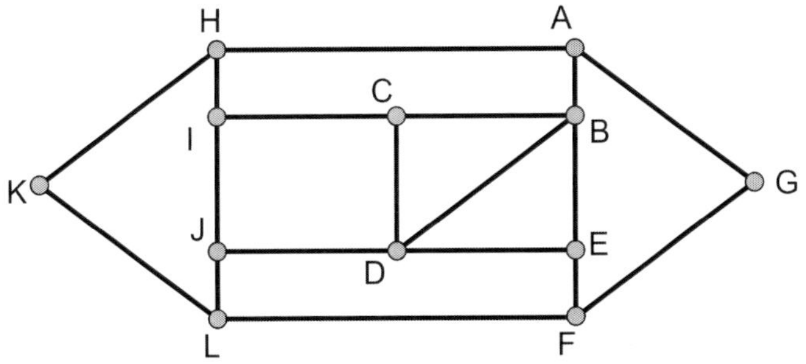

38) Find if the following graph is traversable or not ?

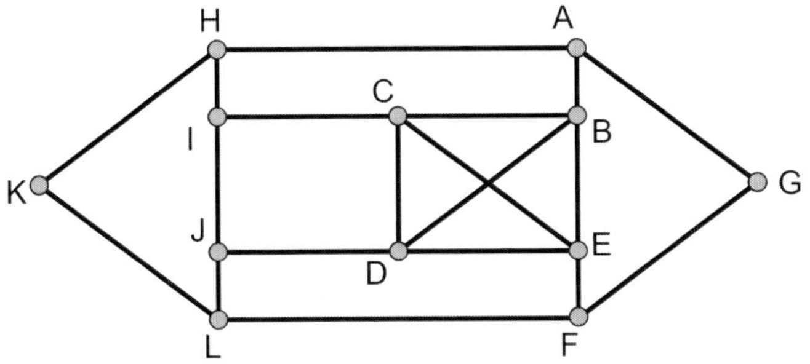

39) Find if the following graph is traversable or not ?

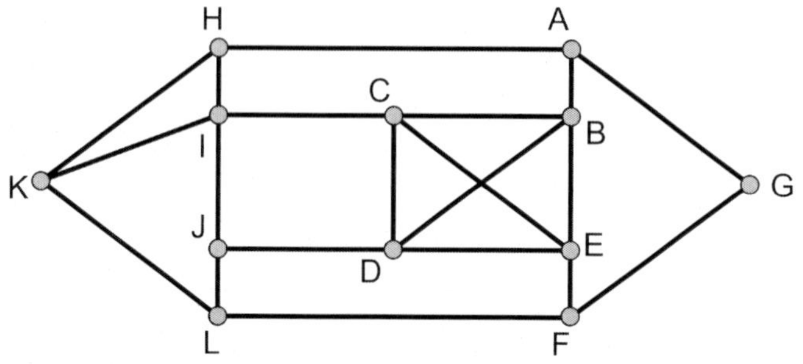

40) Find if the following graph is traversable or not ?

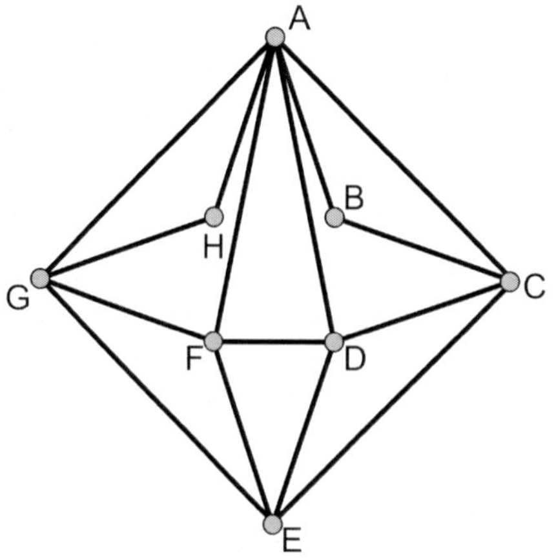

41) Find if the following graph is traversable or not ?

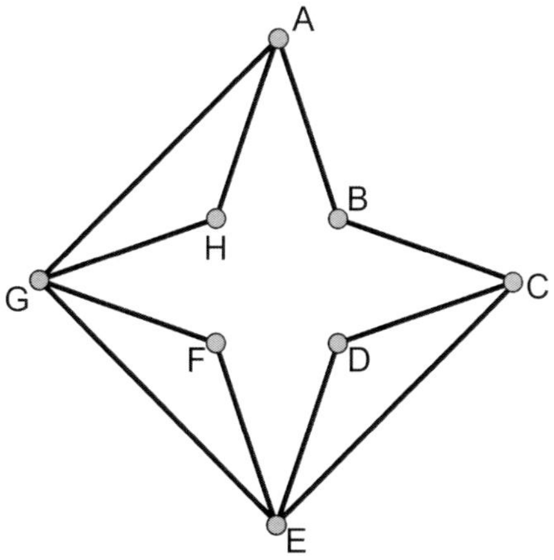

42) Find if the following graph is traversable or not ?

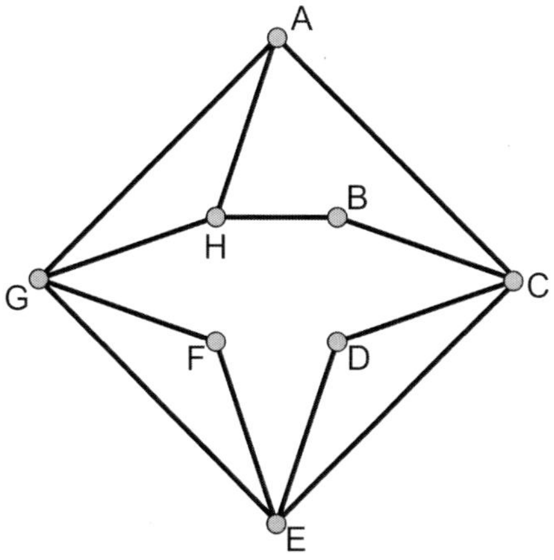

43) Find if the following graph is traversable or not ?

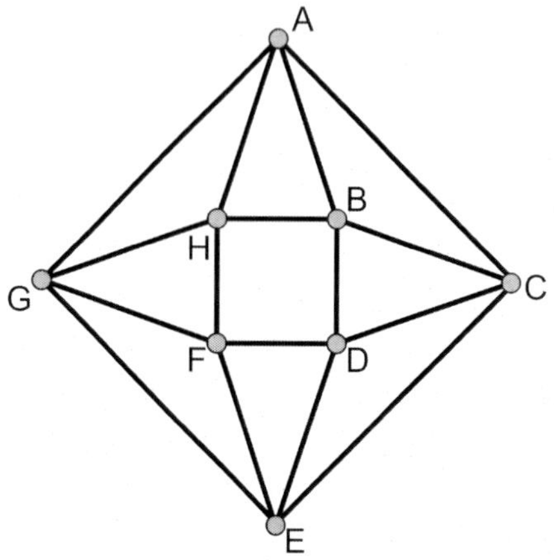

44) Find if the following graph is traversable or not ?

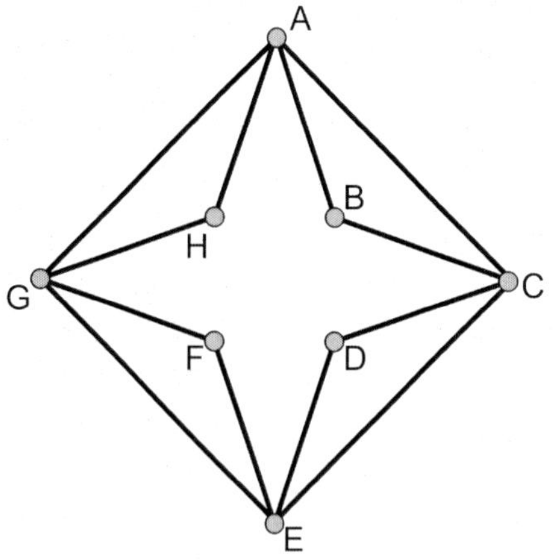

45) Draw an undirected file with the vertices A,B,C,D,E,F,G,H,I,J,K,L and edges AB , BG , AG , GE , GF , EF , EB , HK , HI , IK , IJ , JK , KL , LJ , HA , IC , CB , JD , DE , LF

46) Draw an undirected file with the vertices A,B,C,D,E,F,G,H,I,J,K,L and edges AB , BG , AG , GE , GF , EF , EB , HK , HI , IK , IJ , JK , KL , LJ , HA , IC , CB , JD , DE , LF , IE

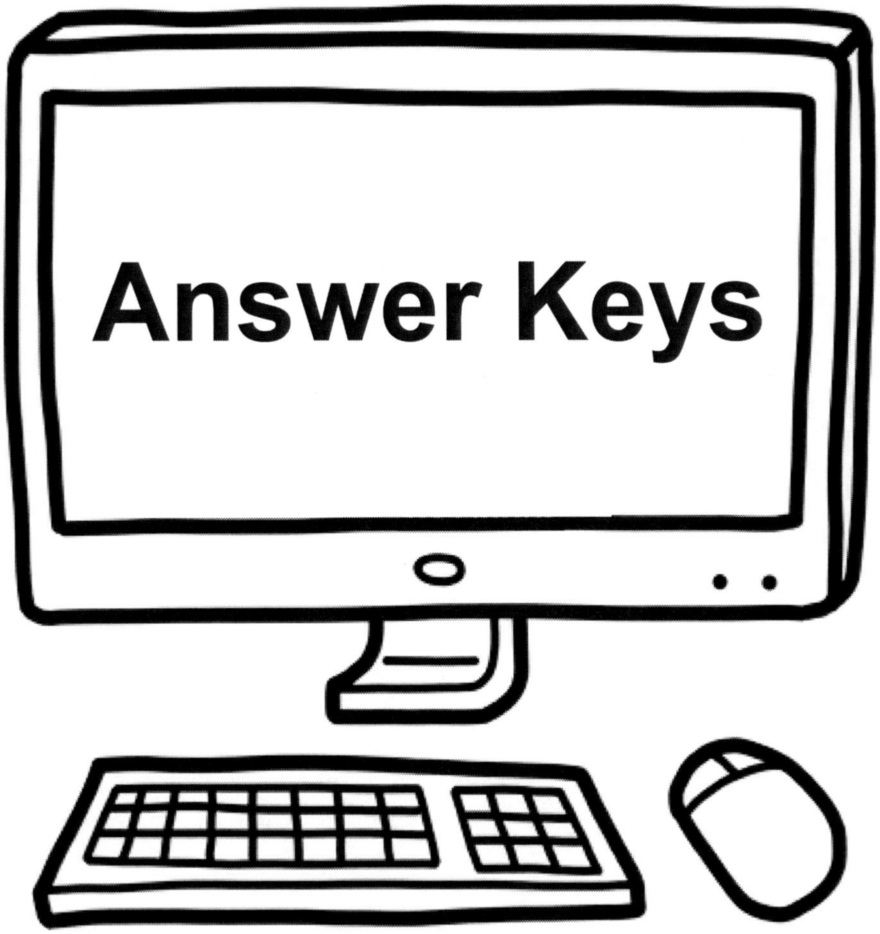
Answer Keys

1. Convert 148_8 to base 10.

148_8 = 1*64 + 4*8 + 8*1 = 64 + 32 + 8 = 104_{10}

2. Convert 109_8 to base 10.

109_8 = 1*64 + 0*8 + 9*1 = 64 + 0 + 9 = 73_{10}

3. Convert 287_8 to base 10.

287_8 = 2*64 + 8*8 + 7*1 = 128 + 64 + 7 = 199_{10}

4. Convert 427_8 to base 10.

427_8 = 4*64 + 2*8 + 7*1 = 256 + 16 + 7 = 279_{10}

5. Convert 538_8 to base 10.

538_8 = 5*64 + 3*8 + 8*1 = 320 + 24 + 8 = 352_{10}

6. Convert 649_8 to base 10.

649_8 = 6*64 + 4*8 + 9*1 = 384 + 32 + 9 = 425_{10}

7. Convert 1009_8 to base 10.

1009_8 = 1*512 + 0*64 + 0*8 + 9*1 = 512 + 0 + 0 + 9
= 521_{10}

8. Convert 2281_8 to base 10.

2281_8 = 2*512 + 2*64 + 8*8 + 1*1 = 1024 + 128 + 64 + 1
= 1217_{10}

9. Convert 10155_8 to base 10.

10155_8 = 1*4096 + 0*512 + 1*64 + 5*8 + 5*1
= 4096 + 0 + 64 + 40 + 5
= 4205_{10}

10. Convert 31189_8 to base 10.

31189_8 = 3*512 + 1*128 + 1*64 + 8*8 + 9*1
= 1536 + 128 + 64 + 64 + 9
= 1801_{10}

11. Convert 2250_8 to base 10. .

2250_8 = 2*512 + 2*64 + 5*8 + 0*1
= 1024 + 128 + 40 + 0
= 1192_{10}

12. Convert $2AF_{16}$ to base 10. .

$2AF_{16}$ = 2*256 + 10*16 + 15*1
= 512 + 160 + 15
= 687_{10}

13. Convert $2AA_{16}$ to base 10. .

$2AA_{16}$ = 2*256 + 10*16 + 10*1
= 512 + 160 + 10
= 682_{10}

14. Convert $5CD_{16}$ to base 10. .

$5CD_{16}$ = 5*256 + 12*16 + 13*1
= 1280 + 192 + 13
= 1485_{10}

15. Convert $2DE_{16}$ to base 10.

$2DE_{16}$ = 2*256 + 13*16 + 14*1
= 512 + 208 + 14
= 734_{10}

16. Convert $A6C_{16}$ to base 10.

A6C = 10*256 + 6*16 + 12*1
= 2560 + 96 + 12
= 2668_{10}

17. Convert $5BF_{16}$ to base 10.

$5BF_{16}$ = 5*256 + 11*16 + 15*1
= 1280 + 176 + 15
= 1471_{10}

18. Convert ACB_{16} to base 10. .

ACB_{16} = 10*256 + 12*16 + 11*1
= 2560 + 192 + 11
= 2763_{10}

19. Convert $BF9_{16}$ to base 10.

$B9F_{16}$ = 11*256 + 9*16 + 15*1
= 2816 + 144 + 15
= 2975_{10}

20. Convert $F8D_{16}$ to base 10. .

$F8D_{16}$ = 15*256 + 8*16 + 13*1
= 3840 + 128 + 13
= 3981_{10}

21. Convert $FE7_{16}$ to base 10.

$FE7_{16}$ = 15*256 + 14*16 + 7*1
= 3840 + 224 + 7
= 4071_{10}

22. Convert the following base 8 number to base 16:
30521_8

$30521_8 = 011\ 000\ 101\ 010\ 001_2$
$= 011\ 0001\ 0101\ 0001_2$
$= 3151_{16}$

23. Convert the following base 8 number to base 16:
75346_8

$75346_8 = 111\ 101\ 011\ 100\ 110_2$
$=111\ 1010\ 1110\ 0110_2$
$= 7AE6_{16}$

24. Convert the following base 8 number to base 16:
51324_8

$51324_8 = 101\ 001\ 011\ 010\ 100_2$
$=101\ 0010\ 1101\ 0100_2$
$= 52D4_{16}$

25. Convert the following base 8 number to base 16:
27105_8

$27105_8 = 010\ 111\ 001\ 000\ 101_2$
$= 010\ 1110\ 0100\ 0101_2$
$= 2E45_{16}$

26. Convert the following base 8 number to base 16:

656565_8

656565_8 = 110 101 110 101 110 101_2

= 11 0101 1101 0111 0101_2

= $35D75_{16}$

27. Convert the following base 8 number to base 16:

135206_8

135206_8 = 001 011 101 010 000 110_2

= 00 1011 1010 1000 0110_2

= $0BA86_{16}$ = $BA86_{16}$

28. Convert the following base 8 number to base 16:

45623_8

45623_8 = 100 101 110 010 011_2

= 100 1011 1001 0011_2

= $4B93_{16}$

29. Convert the following base 8 number to base 16:

727272_8

727272_8 = 111 010 111 010 111 010_2

= 11 1010 1110 1011 1010_2

= $3AEBA_{16}$

30. Convert the following base 8 number to base 16:

141414_8

141414_8 = 001 100 001 100 001 100_2

= 00 1100 0011 0000 1100_2

= $0C30C_{16}$ = $C30C_{16}$

31. Convert the following base 8 number to base 16:

5055055_8

$5055055_8 = 101\ 000\ 101\ 101\ 000\ 101\ 101_2$

$= 1\ 0100\ 0101\ 1010\ 0010\ 1101_2$

$= 145A2D_{16}$

32. Convert $C40DE_{16}$ to Octal.

$C40DE_{16} = 1100\ 0100\ 0000\ 1101\ 1110_2$

$= 11\ 000\ 100\ 000\ 011\ 011\ 110_2$

$= 3040336_8$

33. Convert $BE6FA_{16}$ to Octal. .

$BE6FA_{16} = 1011\ 1110\ 0110\ 1111\ 1010_2$

$= 10\ 111\ 110\ 011\ 011\ 111\ 010_2$

$= 2763372_8$

34. Convert $FDABC_{16}$ to Octal

$FDABC_{16} = 1111\ 1101\ 1010\ 1011\ 1100_2$

$= 11\ 111\ 101\ 101\ 010\ 111\ 100$

$= 3755274_8$

35. Convert $A798C_{16}$ to Octal

$A798C_{16} = 1010\ 0111\ 1001\ 1000\ 1100_2$

$= 10\ 100\ 111\ 100\ 110\ 001\ 100_2$

$= 2474614_8$

36. Convert $DFE5B_{16}$ to Octal

$DFE5B_{16} = 1101\ 1111\ 1110\ 0101\ 1011_2$

$= 11\ 011\ 111\ 111\ 001\ 011\ 011_2$

$= 3377133_8$

37. Convert $EEEDA_{16}$ to Octal

$EEEDA_{16} = 1110\ 1110\ 1110\ 1101\ 1010_2$

$= 11\ 101\ 110\ 111\ 011\ 011\ 010_2$

$= 3567332_8$

38. Convert $C71F8E_{16}$ to Octal

$C71F8E_{16} = 1100\ 0111\ 0001\ 1111\ 1000\ 1110_2$

$= 110\ 001\ 110\ 001\ 111\ 110\ 001\ 110_2$

$= 61617616_8$

39. Convert $DD6BB_{16}$ to Octal

$DD6BB_{16} = 1101\ 1101\ 0110\ 1011\ 1011_2$

$= 11\ 011\ 101\ 011\ 010\ 111\ 011_2$

$= 3353273_8$

40. Convert $BB4FC_{16}$ to Octal

$BB4FC_{16} = 1011\ 1011\ 0100\ 1111\ 1100_2$

$= 10\ 111\ 011\ 010\ 011\ 111\ 100_2$

$= 2732374_8$

41. Convert $AEAED_{16}$ to Octal

$AEAED_{16} = 1010\ 1110\ 1010\ 1110\ 1101_2$

$= 10\ 101\ 110\ 101\ 011\ 101\ 101_2$

$= 2565355_8$

42. How many more 1's are there in the binary representation of $9F4_{16}$ than in the binary representation of $B18_{16}$?

$9F4_{16} = 1001\ 1111\ 0100_2$

$B18_{16} = 1011\ 0001\ 1000_2$

There are 2 more 1's in $9F4_{16}$

43. How many more 1's are there in the binary representa on of $CC97_{16}$ than in the binary representation of $19AD_{16}$?

$CC97_{16} = 1100\ 1100\ 1001\ 0111_2$

$19AD_{16} = 0001\ 1001\ 1010\ 1101_2$

There is 1 more 1 in $CC97_{16}$

44. How many more 1's are there in the binary representation of $77F7_{16}$ than in the binary representation of $5DA2_{16}$?

$77F7_{16} = 0111\ 0111\ 1111\ 0111_2$

$5DA2_{16} = 0101\ 1101\ 1010\ 0010_2$

There are five more 1's in $77F7_{16}$

45. How many more 1's are there in the binary representation of $B97F_{16}$ than in the binary representation of $3D85_{16}$?

$B97F_{16} = 1011\ 1001\ 0111\ 1111_2$

$3D85_{16} = 0011\ 1101\ 1000\ 0101_2$

There are four more 1's in $B97F_{16}$

46. How many more 1's are there in the binary representation of $55DC_{16}$ than in the binary representation of $92EA_{16}$?

$55DC_{16} = 0101\ 0101\ 1101\ 1100_2$

$92EA_{16} = 1001\ 0010\ 1110\ 1010_2$

There is one more 1 in $55DC_{16}$

47. Express the sum in hexadecimal:

$81A_{16} + 5E4_{16}$

$$\begin{array}{r} 81A_{16} \\ +\,5E4_{16} \\ \hline DFE_{16} \end{array}$$

A + 4 = 10 + 4 = 14 = E
1 + E = 1 + 14 = 15 = F
8 + 5 = 13 = D

48. Express the sum in hexadecimal:

$F9D_{16} + DEB_{16}$

$$\begin{array}{r} F9D_{16} \\ +\,DEB_{16} \\ \hline 1D88_{16} \end{array}$$

D + B = 13 + 11 = 24
24 - 16 = 8 (1 carry over)

1 + E + 9 = 1 + 14 + 9 = 24
24 - 16 = 8 (1 carry over)

F + D + 1 = 15 + 13 + 1 = 29
29 -16 =13 = 1D

49. Express the sum in hexadecimal:

$6CD_{16} + 3EE_{16}$

$$\begin{array}{r} 6CD_{16} \\ +\,3EE_{16} \\ \hline ABB_{16} \end{array}$$

D + E = 13 + 14 = 27
27 - 16 = 11 (1 carry over) = B

1 + C + E = 1 + 12 + 14 = 27
27 - 16 = 11 (1 carry over) = B
1 + 6 + 3 = 10 = A

50. Express the sum in hexadecimal:

$EA9_{16} + DF4_{16}$

$$\begin{array}{r} EA9_{16} \\ +\ DF4_{16} \\ \hline IC9D_{16} \end{array}$$

9 + 4 = 13 = D

A + F = 10 + 15 = 25
25 - 16 = 9 (1 carry over)

1 + E + D = 1 + 14 + 13 = 28 28 - 16 = 12 =

C (1 carry over)

51. Express the sum in hexadecimal:

$BFC_{16} + 3DD_{16}$

$$\begin{array}{r} BFC_{16} \\ +\ 3DD_{16} \\ \hline FD9_{16} \end{array}$$

C + D = 12 + 13 = 25
25 - 16 = 9 (1 carry over)

1 + D + F = 1 + 13 + 15 = 29
29 - 16 = 13 (1 carry over) =D

3 + B + 1 = 3 + 11 +1 = 15 = F

52. Simplify the below in hexadecimal:

$9AD_{16} - 2CA_{16}$

$$\begin{array}{r} FAD_{16} \\ -\ 2CA_{16} \\ \hline CE3_{16} \end{array}$$

D - A = 13 - 10 = 3
A - C = 10 - 12 = 16 + 10 -12 (Bring 1 borrow ; 1 =16)
= 14 = E
F - 2 -1 (DONT' FORGET The borrow given earlier)
= 15 - 2 - 1 = 12 = C

53. Simplify the below in hexadecimal:

EAD_{16} - $1FC_{16}$

$$\begin{array}{r} EAD_{16} \\ -\ 1FC_{16} \\ \hline CB1_{16} \end{array}$$

D - C = 13 - 12 = 1

A - F = 10 - 15 = 10 -15 + 16 (Bring 1 borrow ; 1 =16)
= 11 = B

E - 1 - 1 (DONT' FORGET The borrow given earlier)
= 14 - 1 - 1 = 12 = C

54. Simplify the below in hexadecimal:

$9E7_{16}$ - $2BB_{16}$

$$\begin{array}{r} 9E7_{16} \\ -\ 2BB_{16} \\ \hline 73C_{16} \end{array}$$

7 - B = 7 - 11 = 16 + 7 - 11 (Bring 1 borrow ; 1 = 16)
= 12 = C

E - B - 1(remember the borrow given)
= 15 - 11 - 1 = 3

9 - 2 = 9 - 2 = 7

55. Simplify the below in hexadecimal:

$DE3_{16}$ - $5CE_{16}$

$$\begin{array}{r} DE3_{16} \\ -\ 5CE_{16} \\ \hline 815_{16} \end{array}$$

3 - E = 3 - 14 (Bring 1 borrow ; 1 = 16)
= 16 + 3 - 14 = 5

E - C (remember the borrow given)
= 14 - 12 - 1 = 1

D - 5 = 13 - 5
= 8

56. Simplify the below in hexadecimal:

$AC9D_{16} - 1EF7_{16}$

$$\begin{array}{r} AC9D_{16} \\ -\ 1EF7_{16} \\ \hline 8DA6_{16} \end{array}$$

D - 7 = 13 - 7 = 6

9 - F (Bring 1 borrow ; 1 = 16)
= 9 - 15 + 16
= 10 = A

C - E = 12 - 14 - 1 (remember the borrow given)
= 12 - 14 - 1 + 16 (Bring 1 borrow ; 1 = 16= A)
= 13 = D

A - 1 = 10 - 1 - 1 = 8 (remember the borrow given)

57. Express the sum in octa decimal:

$651_8 + 339_8$

$$\begin{array}{r} 651_8 \\ -\ 339_8 \\ \hline 1212_8 \end{array}$$

9 + 1 = 10
10 - 8 = 2 (one carry over 1 = 8)
1 + 3 + 5 = 9
9 - 8 = 1 (one carry over 1 = 8)
6 + 3 + 1 = 10
10 - 8 = 2 (one carry over 1 = 8

58. Express the sum in octa decimal:

$9905_8 + 5136_8$

$$\begin{array}{r} 9905_8 \\ +\,5136_8 \\ \hline 17243_8 \end{array}$$

5 + 6 = 11

11 - 8 = 3 (one carry over 1 = 8)

1 + 0 + 3 = 4

9 + 1 = 10

10 - 8 = 2 (one carry over 1 = 8)

5 + 9 + 1 = 15

15 - 8 = 7 (one carry over 1 = 8)

59. Express the sum in octa decimal:

$7163_8 + 4509_8$

$$\begin{array}{r} 7163_8 \\ +\,4509_8 \\ \hline 13674_8 \end{array}$$

3 + 9 = 12

12 - 8 = 4 (one carry over 1 = 8)

1 + 6 + 0 = 7

5 + 1 = 6

4 + 7 = 11

11 - 8 = 3 (one carry over 1 = 8)

60. Simplify in octa decimal:

$7777_8 - 2895_8$

$$\begin{array}{r} 7777_8 \\ -\,2895_8 \\ \hline 4662_8 \end{array}$$

7 - 5 = 2

7 - 9 = 7 - 9 + 8 (one borrow 1 = 8)

= 6

7 - 8 - 1(remember one borrow was given ;1 = 8)

= 7 - 8 - 1 + 8 (one borrow 1 = 8)

= 6

7 - 2 -1 (remember one borrow was given ;1 = 8)

= 7 - 3 = 4

61. Simplify in octa decimal:

$7657_8 - 3577_8$

$$
\begin{array}{r}
7657_8 \\
+\,3577_8 \\
\hline
2060_8 \\
\hline
\end{array}
$$

7 - 7 = 0
5 - 7 = 5 - 7 + 8 (one borrow 1 = 8)
= 6
6 - 5 -1 (remember one borrow was given ;1 = 8)
= 6 - 6 = 0
7 - 3 = 2

62. Which is larger from the below ,Order them from least to greatest

(A) 427_8 (B) $2AF_{16}$ (C) 1009_8

1009_8 = 1*512 + 0*64 + 0*8 + 9*1 = 512 + 0 + 0 + 9
= 521_{10}

427_8 = 4*64 + 2*8 + 7*1 = 256 + 16 + 7 = 279_{10}

$2AF_{16}$= 2*256 + 10*16 + 15*1
= 512 + 160 + 15
= 687_{10}

Largest is $2AF_{16}$

Least to greatest 427_8 ,1009_8 , $2AF_{16}$

63. Which is larger from the below ,Order them from least to greatest

(A) $2AA_{16}$ (B) $A6C_{16}$ (C) $5BF_{16}$

$2AA_{16}$ = 2*256 + 10*16 + 10*1
= 512 + 160 + 10
= 682_{10}

$A6C_{16}$ = 10*256 + 6*16 + 12*1
= 2560 + 96 + 12 =
= 2668_{10}

$5BF_{16}$ = 5*256 + 11*16 + 15*1
= 1280 + 176 + 15
= 1471_{10}

Largest is $A6C_{16}$

Least to greatest $2AA_{16}$, $5BF_{16}$, $A6C_{16}$

64. Which is larger from the below ,Order them from least to greatest

(A) ACB_{16} (B) 3189_8C) $B9F_{16}$

ACB_{16} = 10*256 + 12*16 + 11*1
= 2560 + 192 + 11
= 2763_{10}

$B9F_{16}$ = 11*256 + 9*16 + 15*1
= 2816 + 144 + 15
= 2975_{10}

3189_8 = 3*512 + 1*64 + 8*8 + 9*1
= 1536 + 64 + 64 + 9
= 1673_{10}

Largest is $B9F_{16}$
Least to greatest 3189_8 , ACB_{16}, $B9F_{16}$

65. Which is larger from the below , Order them from least to greatest

(A) 2281_8 (B) $2DE_{16}$ (C) $5CD_{16}$

2281_8 = 2*512 + 2*64 + 8*8 + 1*1 = 1024 + 128 + 64 + 1
= 1217_{10}

$2DE_{16}$ = 2*256 + 13*16 + 14*1
= 512 + 208 + 14
= 734_{10}

$5CD_{16}$ = 5*256 + 12*16 + 13*1
= 1280 + 192 + 13
= 1485_{10}

Largest is $5CD_{16}$
Greatest to least $5CD_{16}$, 2281_8 , $2DE_{16}$

Greatest to least :

2DE16 , 22818 , 5CD16

66. Which is larger from the below, Order them from least to greatest

(A) $F8D_{16}$ (B) $FE7_{16}$ (C) 1155_8

$F8D_{16}$ = 15*256 + 8*16 + 13*1
= 3840 + 128 + 13
= 3981_{10}

$FE7_{16}$ = 15*256 + 14*16 + 7*1

= 3840 + 224 + 7
= 4071_{10}

1155_8 = 1*512 + 1*64 + 5*8 + 5*1
= 512 + 64 + 40 + 5
= 621_{10}

Largest is $FE7_{16}$
Greatest to least $FE7_{16}$, $F8D_{16}$, 1155_8

67. What is the decimal value of the green component in the RGB code?

$DC143C_{16}$

Green Component: $14_{16} = 1 * 16 + 4 * 1 = 20$

68. Is the decimal value of the blue and green component in the RGB code given below are equal if so what are the values?

$F08080_{16}$

Green Component: $80_{16} = 8 * 16 + 0 * 1 = 128$

Blue Component: $80_{16} = 8 * 16 + 0 * 1 = 128$

69. What are the decimal values of the Red, Green and Blue components in the RGB code?

$ADFF2F_{16}$

Red Component: AD_{16} = 10 *16 + 13 *1 = 173

Green Component: FF_{16} = 15 *16 + 15 *1 = 255

Blue Component: $2F_{16}$ = 2 *16 + 15 *1 = 47

70. Is the decimal value of the blue and green component in the RGB code given below are equal ? Which one is greater and by what value?

$20B2AA_{16}$

Green Component: $B2_{16} = 11 * 16 + 2 * 1 = 178$

Blue Component: $AA_{16} = 10 * 16 + 10 * 1 = 170$

Green component is more by 8

71. Is the decimal value of the red and green component in the RGB code given below are equal ?

Which one is greater and by what value?

$DA70D6_{16}$

Red Component: $DA_{16} = 13 * 16 + 10 * 1 = 218$

Green Component: $70_{16} = 7 * 16 + 0 * 1 = 112$

Red component is more by 106

1. Evaluate the following postfix expression. Note that all numbers are single digit.

8 4 / 5 4 * 2 − +

SOLUTION:

= (8 /4) ((5 * 4) 2 −) + = 2 (20 − 2) +
= 3 18 + = 3 + 18 = 21

2. Evaluate the following postfix expression. Note that all numbers are not single digit.

10 / 11 + 45 * 4 - +

SOLUTION:

(10 / (1 + 1))(4*5 − 4) +
= 5 (20 - 4) + = 5 16 + = 5 + 16 = 21

3. Evaluate the following postfix expression. Note that all numbers are not single digit.

10 5 / 9 3 * 3 − +

SOLUTION:

= (10 /5) ((9 * 3) 3 −) +
= 2 (27− 3) +
= 2 24 + = 2 + 24
= 26

4. Evaluate the following postfix expression. Note that all numbers are not single digit.

44 4 / 10 2 * 7 − +

SOLUTION:

= (44 / 4) ((10 * 2) 7 −) +

= 11 (20− 7) +

= 11 13 + = 11 + 13

= 24

5. Evaluate the following postfix expression. Note that all numbers are not single digit.

36 6 / 4 4 * 9 − +

SOLUTION:

= (36 /6) ((4 * 4) 9 −) +
= 6 (16 − 9) +
= 6 7 + = 6 + 7
= 13

6. Evaluate the following postfix expression. Note that all numbers are not one digit.

7 7 * 3 4 + / 14 2 * 3

- **SOLUTION:**

(7*7) / (3 + 4) + 14*2 − 3
= (49 /7) + 28 - 3 = 7 + 25 = 32

7. Evaluate the following postfix expression if all numbers are single digits.

5 2 4 + 2 ^ * 8 4 / 2 * −

SOLUTION:

= 5 (2 + 4) 2 ^ * (8 / 4) 2 * −

= 5 6 2 ^ * 2 2 * −

= 5 (6 ^2) * (2 * 2) −

= 5 36 * 4 − = (5 * 36) 4 − = 180 4 −

= 180 − 4 = 176

8. Evaluate the following postfix expression if all numbers are single digits.

9 1 1 + 4 ^ * 12 6 / 7 * −

SOLUTION:

= 9 (1 + 1) 4 ^ * (12 / 6) 7 * −
= 9 2 4 ^ * 2 7 * −
= 3(2 ^4) * (2 * 7) −
= 3 16 * 14 − = (3 * 16) 14 −
= 48 14 − = 48 − 14
= 34

9. Evaluate the following postfix expression if all numbers are single digits.

1 1 2 + 2 ^ * 2 2 / 1 * -

SOLUTION:

= 1 (1 + 2) 2 ^ * (2 / 2) 1 * -
= 1 3 2 ^ * 1 1 * -
= 1 (3 ^2) * (1 * 1) -
= 1 9 * 1 - = (1 * 9) 1 -
= 9 1 - = 9 - 1 = 8

10. Evaluate the following postfix expression if all numbers are not single digits.

5 1 5 + 2 ^ * 15 5 / 6 * -

SOLUTION:

= 5 (1 + 5) 2 ^ * (15 / 5) 6 * - = 5
6 2 ^ * 3 6 * -
= 3 (6 ^2) * (3 * 6) -
= 3 36 * 18 - = (3 * 36) 18 - '
= 108 18 - = 108 - 18
= 90

11. Evaluate the following postfix expression if all numbers are not single digits.

7 4 3 - 9 ^ * 39 3 / 2 * +

SOLUTION:

= 7 (4 - 3) 9 ^ * (39 / 3) 2 * + =
7 1 9 ^ * 13 2 * + = 7 (1
^ 9) * (13 * 2) +
= 7 1 * 26 + = (7 * 1) 26 +
= 7 26 + = 7 + 26 = 33

12. Evaluate the following prefix expression. Note that all numbers are single digits in the expression.

+ / + 6 6 6 ^ 4 1

SOLUTION:

=+ / (6 + 6) 6 (4 ^ 1)
= + / 12 6 4 = + (12 / 6) 4
= + 2 4 = 2 + 4
= 6

13. Evaluate the following prefix expression if all numbers are single digits.

− * + 4 6 8 * 4 / 8 4

SOLUTION:

= − * (4 + 6) 8 * 4 (8 / 4)
= − * 10 8 * 4 2 = − (10 * 8) (4 * 2)
= − 80 8 = 80 − 8 = 72

14. Evaluate the following prefix expression. Note that all numbers are single digits in the expression.

- * + 1 2 4 ^ 4 1

SOLUTION:

= - * (1 + 2) 4 (4 ^ 1)
= - * 3 4 4
= - 3 * 4 4
= − 12 4 = 12 - 4
= 8

15. Evaluate the following prefix expression if all numbers are single digits.

− * + 3 1 2 * 4 / 6 3

SOLUTION:

= − * (3 + 1) 2 * 4 (6 / 3)
= − * 4 2 * 4 2 = − (4 * 2) (4 * 2)
= − 8 8 = 8 − 8 = 0

16. Evaluate the following prefix expression. Note that all numbers are single digits in the expresion.

- * + 3 2 5 ^ 3 1

SOLUTION:

= - * (3 + 2) 5 (3 ^ 1)
= - * 5 5 3 = - 5 * 5 3 = − 25 3
= 25 - 3
= 22

17. Evaluate the following prefix expression if all numbers are single digits.

- * + 7 1 5 * 3 / 9 3

SOLUTION:

= − * (7 + 1) 5 * 3 (9 / 3)
= − * 8 5 * 3 3 = − (8 * 5) (3 * 3)
= − 40 9 = 40 − 9 = 31

18. Evaluate the following prefix expression. Note that all numbers are single digits in the expression.

* ^ / 6 3 2 + 8 2

SOLUTION:

= * ^ (6 / 3) 2 (8 + 2)
= * ^ 2 2 10 = * 2 ^ 2 10
= * 4 10 = 4 *10
= 40

19. Evaluate the following prefix expression if all numbers are single digits.

- ^ + 5 5 2 * 5 / 4 2

SOLUTION:

= − ^ (5 + 5) 2 * 5 (4 / 2)
= − ^10 2 * 5 2 = − (10 ^ 2) (5 * 2)
= − 100 10 = 100 − 10 = 90

20. Evaluate the following prefix expression. Note that all numbers are single digits in the expression.

+ / + 3 3 3 ^ 3 1

SOLUTION:

= + / (3 + 3) 3 (3 ^ 1)
= + / 6 3 3 = − (6 / 3) 3
= + 2 3 = 2 + 3
= 5

21. Evaluate the following prefix expression if all numbers are single digits.

− * + 9 3 6 * 4 / 6 2

SOLUTION:

= − * (9 + 3) 6 * 4 (6 / 2)
= − * 12 6 * 4 3 = − (12 * 6) (4 * 3)
= − 72 12 = 72 − 12 = 60

22. Evaluate the following prefix expression if all numbers are not single digits.

+ * / + 20 5 5 ^ 2 / 12 4 35

SOLUTION:

= + * / (20 + 5) 5 ^ 2 (12 / 4) 35
= + * / 25 5 ^ 2 3 20 = + * (25/5) (2 ^ 3) 35
= + * 5 8 35 = + 5 * 8 35 = + 40 35 = 40 + 35 = 75

23. Convert the infix expression to a postfix expression. Perform all operations in the same order as the given expression.

8 * 3 / 6 − 4 ^ 1 - 2 * 3 + 3 ^ 3

SOLUTION:

8 * 3 / 6 − 4 ^ 1 - 2 * 3 + 3 ^ 3
= ((8 3 *) / 6) − (4 1 ^) − (2 3 *) + 3 3 ^
= (8 3 * 6 /) − (4 1 ^) − (2 3 *) + 3 3 ^
= ((8 3 * 6 /) (4 1 ^) −) - (2 3 *) + 3 3 ^
= ((8 3 * 6 /) (4 1 ^) −) (2 3 *) − 3 3 ^ +
= 8 3 * 6 / 4 1 ^ − 2 3 * − 3 3 ^ +

24. Convert the infix expression to a postfix expression. Perform all operations in the same order as the given expression.

(8 + 1) / 9 * 8 / (8 − 2)

SOLUTION:

(8 + 1) / 9 * 8 / (8 − 2)
= (((8 1 +) / 9) * 8) / (8 2 −)
= ((8 1 + 9 /) * 8) / (8 2 −)
= ((8 1 + 9 /) 8 *) / (8 2 −)
= ((8 1 + 9 /) 8 *) (8 2 −) /
= 8 1+ 9 / 8 * 8 2 − /

25. Convert the infix expression to a postfix expression. Perform all operations in the same order as the given expression.

9 − 2 * 7 + 3 * 1 + 5

SOLUTION:

9 − 2 * 7 + 3 * 1 + 5
= 9 − (2 * 7) + (3 * 1) + 5
= 9 − (2 7 *) + (3 1 *) + 5
= (9 (2 7 *) −) + ((3 1 *) 5 +)
= ((9 (2 7 *) −) (3 1 *) 5 +) +
= 9 2 7 * − 3 1 * 5 + +

26. Convert the infix expression to a postfix expression. Perform all operations in the same order as the given expression.

(9 - 1) ^ 4 * (7 + 3) ^ 4 / 5

SOLUTION:

(9 - 1)^4 * (7 + 3)^4 / 5
= (− 9 1)^4 * (+ 7 3)^4 / 5
= (^ (− 9 1) 4) * (^ (+ 7 3) 4) / 5
= (* (^ (− 9 1) 4) (^ (+ 7 3) 4)) / 5
= / (* (^ (− 9 1) 4) (^ (+ 7 3) 4)) 5
= / * ^ − 9 1 4 ^ + 7 3 4 5

27. Convert the infix expression to a prefix expression. Perform all operations in the same order as the given expression.

32 – 9 * 7 + 8 * 5 – 1

SOLUTION:

32 – 9 * 7 + 8 * 5 – 1
= 32 – (9 * 7) + (8 * 5) – 1
= 32 – (* 9 7) + (* 8 5) – 1
= (– 32 (* 9 7)) + (* 8 5) – 1
= (+ (– 32 (* 9 7)) (* 8 5)) – 1
= – (+ (– 32 (* 9 7)) (* 8 5)) 1
= – + – 32 * 9 7 * 8 5 1

28. Convert the infix expression to prefix. Perform all operations in the same order as the expression.

7 / (5 − 4) + (8 + 2)^5 + 8/4

SOLUTION:

7 / (5 − 4) + (8 + 2)^5 + 8/4
= 7 / (− 5 4) + (+ 8 2) ^ 5 +/ 8 4
= (/ 7 (− 5 4)) + (^ (+ 8 2) 5) +/ 8 4
= + (/ 7 (− 5 4)) (^ (+ 8 2) 5) +/ 8 4
= + / 7 − 5 4 ^ + 8 2 5 + / 8 4

29. Convert the infix expression to a prefix expression. Perform all operations in the same order as the given expression.

$10^{(4-2)} * (7+3)^4$

SOLUTION:

$10^{(4-2)} * (7+3)^4$

= 10 ^ (− 4 2) * (+ 7 3) ^ 4

= ^ 10 (− 4 2) * (+ 7 3) ^ 4
= * ^ 10 (− 4 2)(^ (+ 7 3) 4)
= * ^ 10 − 4 2 ^ + 7 3 4

30. Convert the infix expression to a prefix expression. Perform all operations in the same order as the given expression.

$8^{(8-5)} / (6+4)^7$

SOLUTION:

$8^{(8-5)} / (6+4)^7$

= 8 ^ (− 8 5) / (+ 6 4) ^ 7

= ^ 8 (− 8 5) / (+ 6 4) ^ 7
= / ^ 8 (− 8 5)(^ (+ 6 4) 7)
= / ^ 8 − 8 5 ^ + 6 4 7

31. Translate the following postfix expression to prefix if all numbers are single digits.

6 4 + 3 * 6 3 4 + * −

SOLUTION:

6 4 + 3 * 6 3 4 + * −
= (6 + 4) 3 * 6 (3 + 4) * −
= ((6 + 4) * 3) (6 * (3 + 4)) −
= ((6 + 4) * 3) − (6 * (3 + 4))
= ((+ 6 4) * 3) − (6 * (+ 3 4))
= (* (+ 6 4) 3) − (* 6 (+ 3 4))

= − (* (+ 6 4) 3) (* 6 (+ 3 4))
= − * + 6 4 3 * 6 + 3 4

32. Translate the following postfix expression to prefix if all numbers are single digits.

4 1 − 7 4 + 1 * 7 1 4 + * −

SOLUTION:

4 1 − 7 4 + 1 * 7 1 4 + * −
= (4 − 1) (7 + 4) 1 * 7 (1 + 4) * −
= (4 − 1) ((7 + 4) * 1) (7 * (1 + 4)) −
=(4 − 1) ((7 + 4) * 1) − (7 * (1 + 4))
= (− 4 1) ((+ 7 4) * 1) − (7 * (+ 1 4))
= (* (− 4 1) (+ 7 4) 1) − (* 7 (+ 1 4))
= − (* (− 4 1) (+ 7 4) 1) (* 7 (+ 1 4))
= − * − 4 1 + 7 4 1 * 7 + 1 4

33. Translate the following postfix expression to prefix if all numbers are single digits.

3 2 + 1 * 3 2 1 + / −

SOLUTION:

3 2 + 1 * 3 2 1 + / −
= (3 + 2) 1 * 3 (2 + 1) / −
= ((3 + 2) * 1) (3 / (2 + 1)) −
= ((3 + 2) * 1) − (3 / (2 + 1))
= ((+ 3 2) * 1) − (3 / (+ 2 1))
= (* (+ 3 2) 1) − (/ 3 (+ 2 1))
= − (* (+ 3 2) 1) (/ 3 (+ 2 1))
= − * + 3 2 1 / 3 + 2 1

34. Translate the following postfix expression to prefix if all numbers are single digits.

1 1 + 2 * 1 1 2 + * −

SOLUTION:

1 1 + 2 * 1 1 2 + / −

= (1 + 1) 2 * 1 (1 + 2) / −
= ((1 + 1) * 2) (1 / (1 + 2)) −
= ((1 + 1) * 2) − (1 / (1 + 2))
= ((+ 1 1) * 2) − (1 / (+ 1 2))
= (* (+ 1 1) 2) − (/ 1 (+ 1 2))
= − (* (+ 1 1) 2) (/ 1 (+ 1 2))
= − * + 1 1 2 / 1 + 1 2

35. Translate the following prefix expression to postfix. Note that all numbers are single digits in the expression.

* ^ + 3 2 3 + 2 3

SOLUTION:

* ^ + 3 2 3 + 2 3
= * ^ (3 + 2) 3 (2 + 3)
= * ((3 + 2) ^ 3) (2 + 3)
= ((3 + 2) ^ 3) * (2 + 3)
= (((3 2 +) 3 ^) (2 3 +) *)
= 3 2 + 3 ^ 2 3 + *

36. Translate the following prefix expression to postfix. Note that all numbers are single digits in the expression.

/ ^ + 8 4 8 – 4 1

SOLUTION:

/ ^ – 8 4 8 + 4 1
= / ^ (8 – 4) 8 (4 + 1)
= / ((8 – 4) ^ 8) (4 + 1)
= ((8 – 4) ^ 8) / (4 + 1)
= (((8 4 –) 8 ^) (4 1 +) /)
= 8 4 – 8 ^ 4 1 + /

37. Translate the follow postfix expression to prefix. Note that all numbers are single

7 6 – 6 1 + + 4 5 * 3 – +

SOLUTION:

7 6 – 6 1 + + 4 5 * 3 – +
= (7 – 6) (6 + 1) + (4 * 5) 3 – +
= ((7 – 6) + (6 + 1)) ((4 * 5) – 3) +
= ((7 – 6) + (6 + 1)) + ((4 * 5) – 3)
= ((– 7 6) + (+ 6 1)) + ((* 4 5) – 3)
= (+ (– 7 6) (+ 6 1)) + (– (* 4 5) 3)
= + (+ (– 7 6) (+ 6 1)) (– (* 4 5) 3)
= + + – 7 6 + 6 1 – * 4 5 3

38. Translate the follow postfix expression to prefix. Note that all numbers are single digits.

4 3 * 2 4 + + 2 4 * 3 − +

SOLUTION:

4 3 * 2 4 + + 2 4 * 3 − +
= (4 * 3) (2 + 4) + (2 * 4) 3 − +
= ((4 * 3) + (2 + 4)) ((2 * 4) − 3) +
= ((4 * 3) + (2 + 4)) + ((2 * 4) − 3)
= ((* 4 3) + (+ 2 4)) + ((* 2 4) − 3) = (+ (* 4 3) (+ 2 4)) + (− (* 2 4) 3) = + (+ (* 4 3) (+ 2 4)) (− (* 2 4) 3)
= + + * 4 3 + 2 4 − * 2 4

39. Translate the following prefix expression to postfix. Note that all numbers are single digits.

− / + 2 2 1 * * 2 1 2

SOLUTION:

− / + 2 2 1 * * 2 1 2
= − / (2 + 2) 1 * (2 * 1) 2
= − ((2 + 2) / 1) ((2 * 1) * 2)
= ((2 + 2) / 1) − ((2 * 1) * 2)
= ((2 2 +) 1 /) ((2 1 *) 2 *) −
= 2 2 + 1 / 2 1 * 2 * −

40. Translate the following prefix expression to postfix. Note that all numbers are single digits.

+ / + 6 7 6 * * 7 6 7

SOLUTION:

+ / + 6 7 6 * * 7 6 7
= + / (6 + 7) 6 * (7 * 6) 7
= + ((6 + 7) / 6) ((7 * 6) * 7)
= ((6 + 7) / 6) + ((7 * 6) * 7)
= ((6 7 +) 6 /) ((7 6 *) 7 *) +
= 6 7 + 6 / 7 6 * 7 * +

41. Translate the following prefix expression to postfix. Note that all numbers are single digits.

+ / − 8 7 5 * * 5 6 4

SOLUTION:

+ / − 8 7 5 * * 5 6 4
= + / (8 − 7) 5 * (5 * 6) 4
= + ((8 − 7) / 5) ((5 * 6) * 4)

= ((8 − 7) / 5) + ((5 * 6) * 4) = ((8 7 −) 5 /) ((5 6 *) 4 *) +

= 8 7 − 5 / 5 6 * 4 * +

42. Translate the following prefix expression to postfix. Note that all numbers are single digits.

+ / − 9 3 3 * − 9 3 9

SOLUTION:

+ / − 9 3 3 * − 9 3 9
= + / (9 − 3) 3 * (9 − 3) 9
= + ((9 − 3) / 3) ((9 − 3) * 9)
= ((9 − 3) / 3) + ((9 − 3) * 9)
= ((9 3 −) 3 /) ((9 3 −) 9 *) +
= 9 3 − 3 / 9 3 − 9 * +

43. Translate the following prefix expression to postfix. Note that all numbers are single digits.

+ / + 4 5 6 * − 5 6 4

SOLUTION:

+ / + 4 5 6 * − 5 6 4
= + / (4 + 5) 6 * (5 − 6) 4
= + ((4 + 5) / 6) ((5 − 6) * 4)
= ((4 + 5) / 6) + ((5 − 6) * 4) = ((4 5 +) 6 /) ((5 6 −) 4 *) +
= 4 5 + 6 / 5 6 − 4 * +

44. Translate the following prefix expression to postfix. Note that all numbers are single digits.

* / – 9 1 2 * + 9 5 * 8 3

SOLUTION:

* / – 9 1 2 * + 9 5 * 8 3
= * / (9 – 1) 2 * (9 + 5) (8 * 3)
= * ((9 – 1) / 2) ((9 + 5) * (8 * 3))
= ((9 – 1) / 2) * ((9 + 5) * (8 * 3))
= ((9 1 –) 2 /) * ((9 5 +) (8 3 *) *)
= ((9 1 –) 2 /) ((9 5 +) (8 3 *) *) *
= 9 1 - 2 / 9 5 + 8 3 * * *

45. Translate the following prefix expression to postfix. Note that all numbers are single digits

* / – 7 5 4 * – 7 1 + 6 3

SOLUTION:

* / – 7 5 4 * – 7 1 + 6 3

= * / (7 – 5) 4 * (7 – 1) (6 + 3)

= * ((7 – 5) / 4) ((7 – 1) * (6 + 3))

= ((7 – 5) / 4) * ((7 – 1) * (6 + 3))

= ((7 5 –) 4 /) * ((7 1 –) (6 3 +) *)

= ((7 5 –) 4 /) ((7 1 –) (6 3 +) *) *

= 7 5 – 4 / 7 1 – 6 3 + * *

Evaluate the following expression as either TRUE or FALSE:

1. (((12 + 7) <= 18) AND ((13 + 3)>= 4*3)) OR NOT (9 * 1 <= 5 * 0)

   ((FALSE) AND (TRUE)) OR TRUE = FALSE OR TRUE = TRUE

2. ((8/2) >= 5) AND (2 * 0) >= 1 * 1) OR NOT (9/3 > 2 * 2)

   (FALSE AND FALSE) OR NOT(FALSE) = FALSE OR TRUE = TRUE

3. ((8 - 8) <= (8 + 8) OR ((16/4) >= 3 * 4) OR NOT (6 * 2 >= 2 * 6)

   (TRUE OR FALSE) OR FALSE = TRUE OR FALSE = TRUE

4. ((2* (7 + 3) <= 3 * (1 + 7) AND ((5 + 5) >= 4 * 1) OR NOT (63/7 > = 3 * 3)

   (TRUE AND TRUE) OR FALSE = TRUE OR FALSE = TRUE

5. ((3/8 <= 2/7) AND ((1/6) >= 5 * (1/10)) OR NOT (21 - 8 = 2 + 9)

   (FALSE AND FALSE) OR NOT(FALSE) = FALSE OR TRUE = TRUE

6. NOT (4^2 < 2 * 4) OR (45/ 9 ≥ 4/2 AND 55 * 2 < 10 * 5)

   NOT(FALSE) OR (TRUE AND FALSE) = TRUE OR FALSE = TRUE

7. NOT (8^2 < 2^3) OR (100 / 25 ≥ 0 AND 108 * 0 > 100 * 8)

   TRUE OR (TRUE AND FALSE) = TRUE OR FALSE = TRUE

8. NOT (54/2 -1 < = 26) OR (99-90 ≥ 100-99 AND 9^2 < 9 *9)

   TRUE OR (TRUE AND FALSE) = TRUE OR FALSE = TRUE

Evaluate the following expression as either TRUE or FALSE:

9. NOT (11^2 < 2 * 11) OR (55-50 ≥ 3 * 7 AND 130/ 2 > 13 * 5)

   TRUE OR (FALSE AND FALSE) = TRUE OR FALSE = TRUE

10. NOT (2 * (30 + 25) < = 55 + 55) OR (((69 + 23) + 10 = 69 + (23 + 10)) AND ((14*12) * 16 <= 14* (12 *16))

    FALSE OR (TRUE AND TRUE) = FALSE OR TRUE = TRUE

11. How many ordered pairs make the following statement FALSE?

A	B	A OR B	A AND B	NOT A	NOT B	NOT (A OR B)	NOT (A AND B)	NOT (A OR B) OR NOT(A AND B)	(A OR B) OR (NOT (A OR B) OR NOT (A AND B))
1	1	1	1	0	0	0	0	0	1
0	1	1	0	1	0	0	1	1	1
1	0	1	0	0	1	0	1	1	1
0	0	0	0	1	1	1	1	1	1

Zero ordered pairs make the statement FALSE

12. NOT A OR (A OR (A OR B))

A	B	NOT A	A OR B	A OR (A OR B)	NOT A OR (A OR (A OR B))
1	1	0	1	1	1
0	1	1	1	1	1
1	0	0	1	1	1
0	0	1	0	0	1

Evaluate the following expression as either TRUE or FALSE:

13. NOT A OR (A OR (A AND B))

A	B	NOT A	A AND B	A OR (A AND B)	NOT A OR (A OR (A AND B))
1	1	0	1	1	1
0	1	1	0	0	1
1	0	0	0	1	1
0	0	1	0	0	1

TRUE Always

14. A OR NOT (A OR B)

A	B	A OR B	NOT (A OR B)	A OR NOT (A OR B)
1	1	1	0	1
0	1	1	0	0
1	0	1	0	1
0	0	0	1	1

15. A AND NOT (A AND B)

A	B	A AND B	NOT (A AND B)	A AND NOT (A AND B)
1	1	1	0	0
0	1	0	1	0
1	0	0	1	1
0	0	0	1	0

Evaluate the following expression as either TRUE or FALSE:

16. NOT A AND (A AND (A AND B))

A	B	NOT A	A AND B	A AND (A AND B)	(NOT A AND (A AND (A AND B))
1	1	0	1	1	0
0	1	1	0	0	0
1	0	0	0	0	0
0	0	1	0	0	0

17. (A OR NOT NOT B) OR (A OR NOT NOT A) AND (NOT (A AND B))

A	B	NOT A	NOT B	NOT NOT A	NOT NOT B	(A OR NOT NOT B)	(A OR NOT NOT A)	(A OR NOT NOT B) OR (A OR NOT NOT A)	NOT (A AND B)	((A OR NOT NOT B) OR (A OR NOT NOT A)) AND (NOT (A AND B))
1	1	0	0	1	1	1	1	1	0	0
0	1	1	0	0	1	1	0	1	1	1
1	0	0	1	1	0	1	1	1	1	1
0	0	1	1	0	0	0	0	0	1	0

Evaluate the following expression as either TRUE or FALSE:

18. ((B AND NOT NOT B) AND (A OR NOT NOT A)) AND (A AND NOT B)

A	B	NOT A	NOT B	NOT NOT A	NOT NOT B	(B AND NOT NOT B)	(A OR NOT NOT A)	(B AND NOT NOT B) AND (A OR NOT NOT A)	A AND NOT B	((B AND NOT NOT B) AND (A OR NOT NOT A)) AND (A AND NOT B)
1	1	0	0	1	1	1	1	1	0	0
0	1	1	0	0	1	1	0	0	0	0
1	0	0	1	1	0	0	1	0	1	0
0	0	1	1	0	0	0	0	0	0	0

Answer: FALSE

19. (A OR B) AND (B AND A) OR (A AND B)

A	B	A OR B	B AND A	(A OR B) AND (B AND A)	(A OR B) AND (B AND A) AND (A OR B)
1	1	1	1	1	1
0	1	1	0	0	0
1	0	1	0	0	0
0	0	0	0	0	0

Evaluate the following expression as either TRUE or FALSE:

20. ((A AND B) AND (B AND A)) OR (A OR B)

A	B	A OR B	B AND A	(A OR B) AND (B AND A)	(A OR B) OR ((B AND A) AND (A OR B))
1	1	1	1	1	1
0	1	1	0	0	1
1	0	1	0	0	1
0	0	0	0	0	0

21. ((A AND B) OR (B AND NOT B)) AND (B AND NOT NOT B)

A	B	NOT B	NOT NOT B	A AND B	B AND NOT B	B AND NOT NOT B	(A AND B) OR (B AND NOT B)	((A AND B) OR (B AND NOT B)) AND (B AND NOT NOT B)
1	1	0	1	1	0	1	1	1
0	1	0	1	0	0	1	0	0
1	0	1	0	0	0	0	0	0
0	0	1	0	0	0	0	0	0

Evaluate the following expression as either TRUE or FALSE:

22. ((A AND B) AND (B AND NOT B)) OR (A AND NOT NOT A)

A	B	NOT B	NOT NOT B	A AND B	B AND NOT B	A AND NOT NOT A	(A AND B) AND (B AND NOT B)	((A AND B) AND (B AND NOT B)) OR (A AND NOT NOT A)
1	1	0	1	1	0	1	0	1
0	1	0	1	0	0	0	0	0
1	0	1	0	0	0	1	0	1
0	0	1	0	0	0	0	0	0

23. ((A AND B) AND (B AND NOT B)) AND (B AND NOT NOT B)

A	B	NOT B	NOT NOT B	A AND B	B AND NOT B	B AND NOT NOT B	(A AND B) AND (B AND NOT B)	((A AND B) AND (B AND NOT B)) AND (B AND NOT NOT B)
1	1	0	1	1	0	1	0	0
0	1	0	1	0	0	1	0	0
1	0	1	0	0	0	0	0	0
0	0	1	0	0	0	0	0	0

Evaluate the following expression as either TRUE or FALSE:

24. ((A OR B) AND (B OR NOT B)) AND (B OR NOT NOT B)

A	B	A OR B	NOT B	B OR NOT B	B OR NOT NOT B	(A OR B) AND (B OR NOT B)	((A OR B) AND (B OR NOT B)) AND (B OR NOT NOT B)
1	1	1	0	1	1	1	1
0	1	1	0	1	1	1	1
1	0	1	1	1	0	1	0
0	0	0	1	1	0	0	0

25. (NOT ((A AND A) AND (B OR B))) AND NOT (A AND B)

A	B	A AND B	NOT (A AND B)	A AND A	B OR B	(A AND A) AND (B OR B))	NOT ((A AND A) AND (B OR B))	(NOT ((A AND A) AND (B OR B))) AND NOT (A AND B)
1	1	1	0	1	1	1	0	0
0	1	0	1	0	1	0	1	1
1	0	0	1	1	0	0	1	1
0	0	0	1	0	0	0	1	1

Evaluate the following expression as either TRUE or FALSE:

26. (NOT ((A OR A) OR (B OR B))) OR NOT (A OR B)

A	B	A OR A	B OR B	A OR B	NOT (A OR B)	((A OR A) OR (B OR B))	NOT ((A OR A) OR (B OR B))	(NOT ((A OR A) OR (B OR B))) OR NOT (A OR B)
1	1	1	1	1	0	1	0	0
0	1	0	1	1	1	1	0	1
1	0	1	0	1	0	1	0	0
0	0	0	0	0	1	0	1	1

27. (NOT ((A OR A) AND (B OR B))) AND NOT (A OR B)

A	B	A OR A	B OR B	A OR B	NOT (A OR B)	((A OR A) AND (B OR B))	NOT ((A OR A) AND (B OR B))	(NOT ((A OR A) OR (B OR B))) AND NOT (A OR B)
1	1	1	1	1	0	1	0	0
0	1	0	1	1	0	0	1	0
1	0	1	0	1	0	0	1	0
0	0	0	0	0	1	0	1	1

Evaluate the following expression as either TRUE or FALSE:

28. ((NOT A OR A) AND (B OR NOT B)) OR NOT (A OR B)

A	B	NOT A	NOT B	NOT A OR A	B OR NOT B	(NOT A OR A) AND (B OR NOT B)	NOT (A OR B)	((NOT A OR A) AND (B OR NOT B)) OR NOT (A OR B)
1	1	0	0	1	1	1	0	1
0	1	1	0	1	1	1	0	1
1	0	0	1	1	1	1	0	1
0	0	1	1	1	1	1	1	1

29. ((A AND NOT NOT A) OR (A OR NOT NOT A)) AND NOT (A AND B)

A	B	NOT A	NOT B	NOT NOT A	A AND B	A AND NOT NOT A	A OR NOT NOT A	NOT (A AND B)	(A AND NOT NOT A) OR (A OR NOT NOT A)	((A AND NOT NOT A) OR (A OR NOT NOT A)) AND NOT (A AND B)
1	1	0	0	1	1	1	1	0	1	0
0	1	1	0	0	0	0	0	1	0	0
1	0	0	1	1	0	1	1	1	1	1
0	0	1	1	0	0	0	0	1	0	0

Evaluate the following expression as either TRUE or FALSE:

30. ((A OR NOT A) AND (B OR NOT B)) AND (B OR NOT A)

A	B	NOT A	NOT B	A OR NOT A	B OR NOT B	B OR NOT A	(A OR NOT A) AND (B OR NOT B)	((A OR NOT A) AND (B OR NOT B))AND (B OR NOT A)
1	1	0	0	1	1	1	1	1
0	1	1	0	1	1	1	1	1
1	0	0	1	1	1	0	1	0
0	0	1	1	1	1	1	1	1

31. ((B OR NOT A) AND (A OR NOT A)) AND (A OR NOT B)

A	B	NOT A	NOT B	A OR NOT A	B OR NOT A	A OR NOT B	(B OR NOT A) AND (A OR NOT A)	((B OR NOT A) AND (A OR NOT A))AND (A OR NOT B)
1	1	0	0	1	1	1	1	1
0	1	1	0	1	1	0	1	0
1	0	0	1	1	0	1	0	0
0	0	1	1	1	1	1	1	1

Simplify the following Boolean expression:

32. (~A + ~B) (A + B) = (~AA + ~AB + ~BA + ~B.B)

= (0 + ~A B + B A + 0) = (~A + ~B + B + A)

= 1

= TRUE

33. A (~A + B) (~A + ~B) = A (~A~A + ~A~B + B~A + B.~B)

= A (~A + ~A ~B + B ~A + 0)

= A (~A + ~A (B + ~B))

= A .(~A + ~A) = A. ~A = 0 = FALSE

34. ~ (~A + B) (A + ~B) = ~ (~AA + ~A~B + BA + B.~B)

= ~ (0 + ~A ~B + B A + 0)

= ~ (~A + ~B + B + A)

= ~ (1) = 0 = FALSE

35. Is the below statement TRUE or FALSE?

~A . ~B + B (A + ~B) = ~A . ~B + B. A + B. ~B = ~A . ~B + BA = ~A + ~B + A + B = TRUE

36. ~ [A (B + ~B) (A + ~A) + B] = ~ [A(1)(1) + B]
= ~[A + B] = ~(AB)

37. ~ [A (B + ~B) (A + ~A)] + ~B = ~ [A(1)(1]) + ~B

= ~A + ~B
= ~AB

38. ~ [A (B + ~B) (A + ~A)] + B = ~ [A(1)(1)] + B

= ~A + B

39. A.(~A + B) + A ~B + ~ A

= A . ~A + A .B + A. ~B + ~A = 0 + A (B + ~B) + ~A

= A + ~A = 1 = TRUE

40. B[~A(~A + ~B) + A .~B] + A] + ~B = B[[~ A . ~A + ~A . ~B + A . ~B] + A] + ~B

= B[[~A + ~B (~A + A)] + A] + ~B
= B[~A + ~B + A] + ~B
= B [1 + ~B] = B + ~B. B + ~B = 1 + 0 = 1 = TRUE

41. [A (B + ~B) (A + ~A)] + (~~A + ~A)(~A) = [A. (1) (1)] + (A + ~A) (~A)

= A + ~A = 1 =TRUE

42. [~A (B + ~B) (A + ~A)] + (~~A + ~A)(A) = [~A. (1) (1)] + (A + ~A) (A)

= A + ~A = 1 = TRUE

43. A[~A (B + ~B) (A + ~A)] + (~~A)(~A) = A[[~A. (1) (1)] + (~ ~A) (~A)

= A.~A + A.~A = 0 + 0 = 0 = FALSE

44. ~B (~AB + B) + ~A~B = ~B ~A B + ~BB + ~A~B
= 0~A + 0 + ~A~B
= 0 + ~A~B
= ~A~B

45. B (~A~B + ~B) + ~B (B + ~B) + B = B~A ~B + B~B + ~BB + ~B~B + B
= 0 + 0 + 0 + ~B + B
= B + ~B = 1 = TRUE

46. (A + ~ B)(~A + B) + A(~B) + B~A = A(~A) + AB + ~B~A + ~BB + A~B + B~A

= 0 + AB + ~B~A + A~B + B~A
= A (B + ~B) + ~B~A + B~A
= A (1) + ~B~A + B~A
= A + ~A (~B + B)
= A + ~A(1) = A + ~A = 1

USE FORMULAE
{~BB =0 , B + ~B = 1 ,A + A = A , A + ~ A = 1}

1) Green island has 12 visiting places connected by bridges to walk as given below. Jack can go from A to B , B to C , D to C , D to E , F to E , F to G , G to H , J to I , I to H , K to J , A to L , and L to K. Draw a graph showing the map with connections to all the bridges in Green island.

Solution :

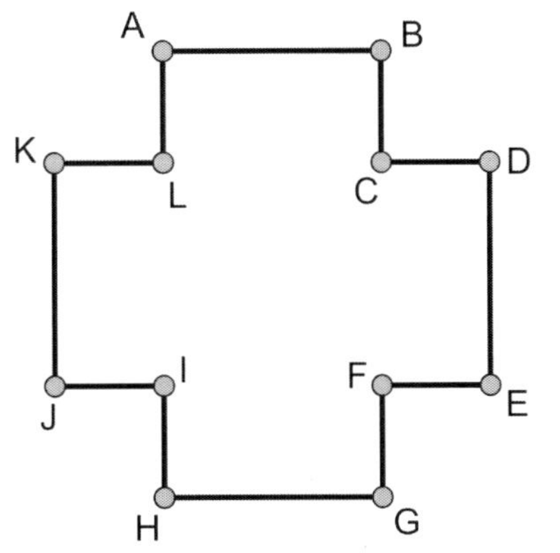

2) Green island has 12 visiting places connected by bridges to walk as given below. Jack can go from A to B , B to C , D to C , D to E , F to E , F to G , G to H , J to I , I to H , K to J , A to L , L to K , and F to L. Draw a graph showing the map with connections to all the bridges in Green island.

Solution :

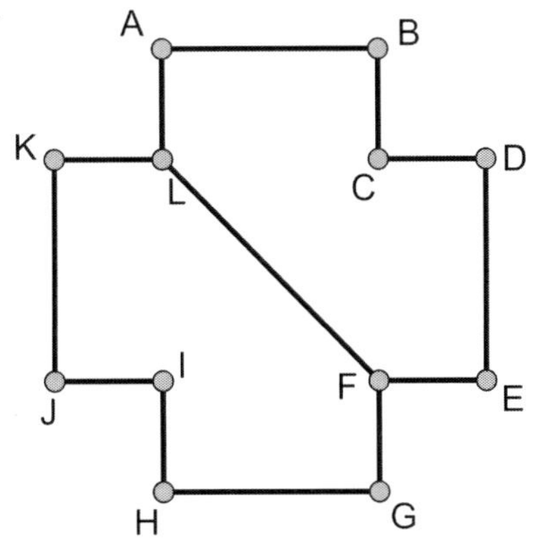

3) Green island has 12 visiting places connected by bridges to walk as given below. Jack can go from A to B , B to C , D to C , D to E , F to E , F to G , G to H , J to I , I to H , K to J , A to L , L to K , C to I and F to L. Draw a graph showing the map with connections to all the bridges in Green island.

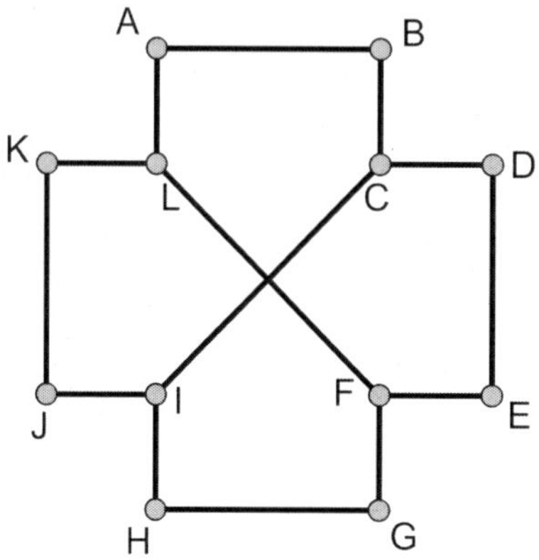

4) Green island has 12 visiting places connected by bridges to walk as given below. Jack can go from A to B , B to C , D to C , D to E , F to E , F to G , G to H , J to I , I to H , K to J , A to L , L to K , C to I , A to K and F to L. Draw a graph showing the map with connections to all the bridges in Green island.

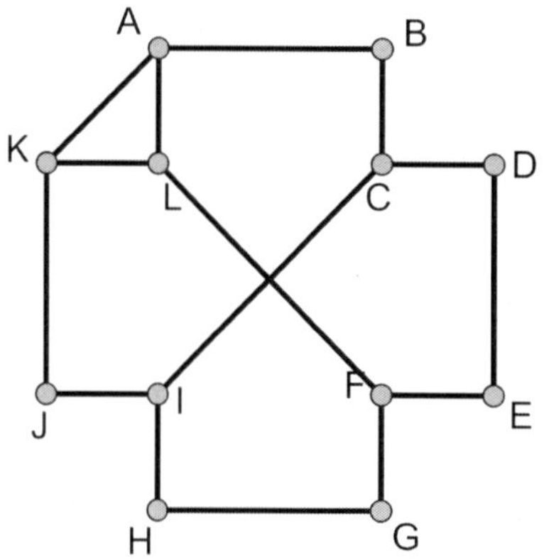

5) Crazy park has 12 rides A , B , C , D , E , F , G , H , I , J , K , L.
The rides are connected to each other by a certain walking path as below.
A to B , B to C , D to C , D to E , F to E , F to G , G to H , J to I , I to H , K to J , A to L , L to K , C to I , A to K , F to L , and D to B
Draw a graph showing the map with the path to all the rides.

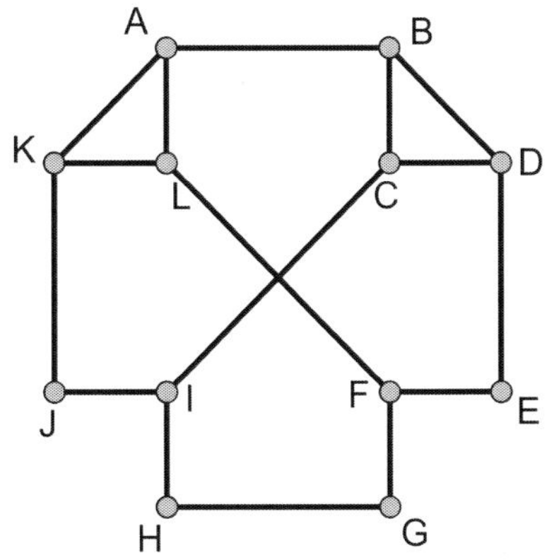

6) Crazy park has 12 rides A , B , C , D , E , F , G , H , I , J , K , L.
The rides are connected to each other by a certain walking path as below.
A to B , B to C , D to C , D to E , F to E , F to G , G to H , J to I , I to H , K to J , A to L , L to K , C to I , A to K , F to L , D to B , J to H , and G to E .
Draw a graph showing the map with the path to all the rides.

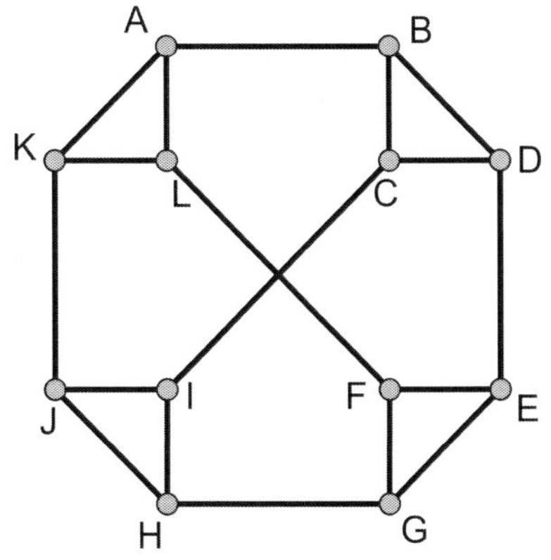

7) Crazy park has 12 rides A , B , C , D , E , F , G , H , I , J , K , L.
The rides are connected to each other by a certain walking path as below.
A to B , B to C , D to C , D to E , F to E , F to G , G to H , J to I , I to H , K to J , A to L , L to K , C to I , A to K , D to B , J to H , and G to E .
Draw a graph showing the map with the path to all the rides.

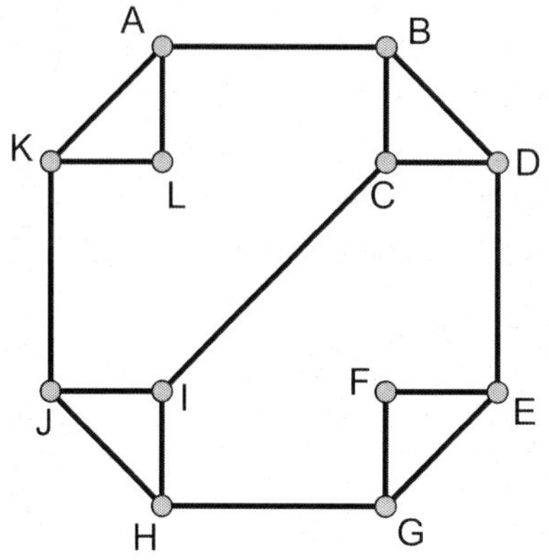

8) Crazy park has 12 rides A , B , C , D , E , F , G , H , I , J , K , L.
The rides are connected to each other by a certain walking path as below.
A to B , B to C , D to C , D to E , F to E , F to G , G to H , J to I , I to H , K to J , A to L , L to K , A to K , D to B , J to H , and G to E .
Draw a graph showing the map with the path to all the rides.

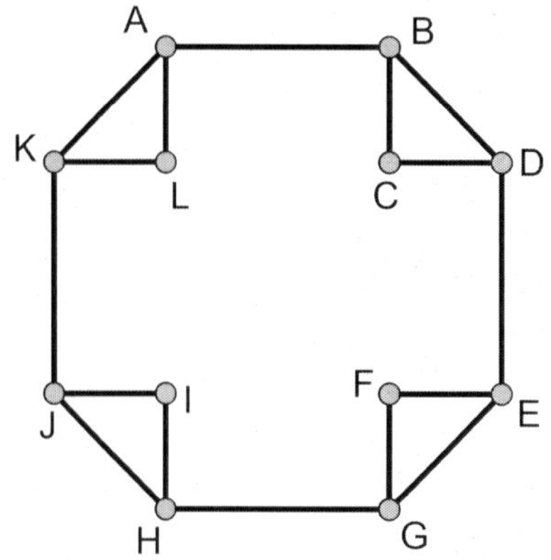

9) Julia has 4 horse stables in her farm. Each stable is connected to the other by a road as given below
A to B , B to C , D to C , D to E , F to E , F to G , G to H , J to I , I to H , K to J , A to L , L to K , A to K , D to B , I to L , J to H , and G to E.
Draw a graph showing the map with connecting the horse stables at Julia's farm.

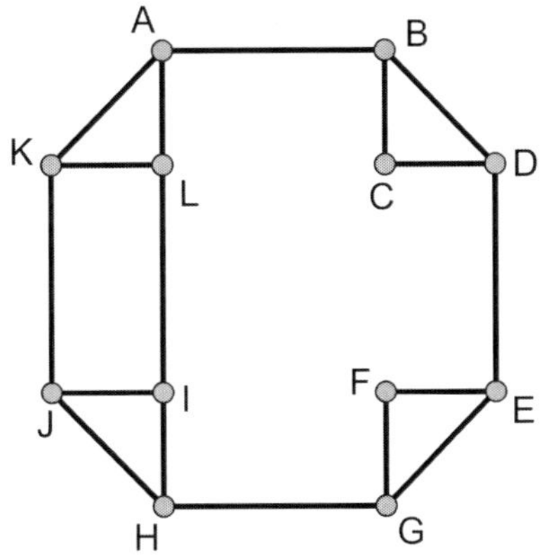

10) Julia has 4 horse stables in her farm. Each stable is connected to the other by a road as given below
A to B , B to C , D to C , D to E , F to E , F to G , G to H , J to I , I to H , K to J , A to L , L to K , A to K , D to B , I to L , C to F , J to H , and G to E.
Draw a graph showing the map with connecting the horse stables at Julia's farm.

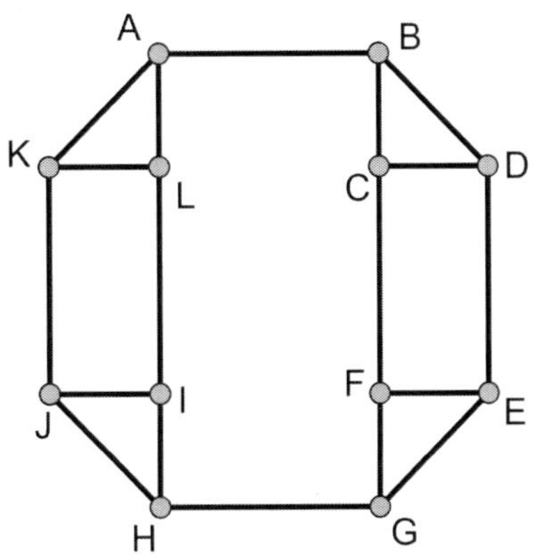

11) Julia has 4 horse stables in her farm. Each stable is connected to the other by a road as given below
A to B , B to C , D to C , D to E , F to E , F to G , G to H , J to I , I to H , K to J , A to L , L to K , A to K , D to B , I to L , L to C , C to F , J to H , and G to E. Draw a graph showing the map with connecting the horse stables at Julia's farm.

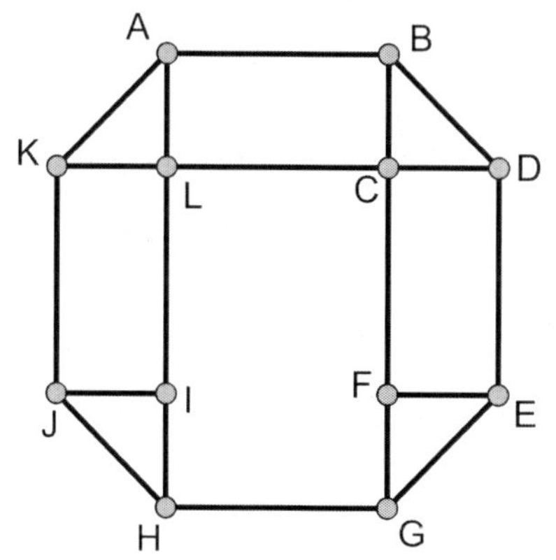

12) Julia has 4 horse stables in her farm. Each stable is connected to the other by a road as given below
A to B , B to C , D to C , D to E , F to E , F to G , G to H , J to I , I to H , K to J , A to L , L to K , A to K , D to B , I to L , L to C , C to F , I to F , J to H , and G to E. Draw a graph showing the map with connecting the horse stables at Julia's farm.

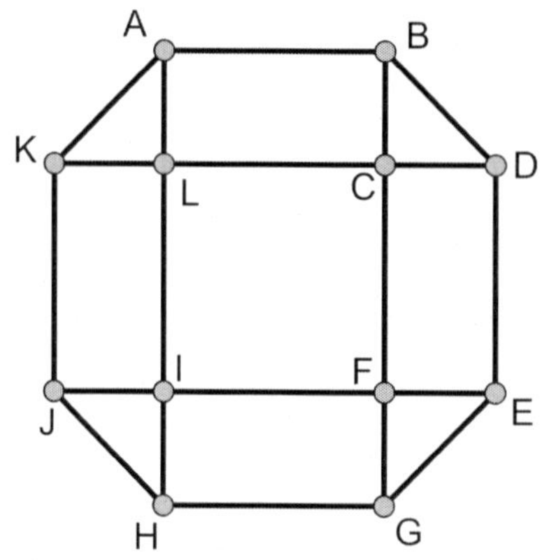

13) Draw an undirected graph with the vertices A,B,C,D,E,F,G,H,I,J,K , and L. Edges as AB , BC , CD , BD , CF , FE , ED , EG , GF , FI , HG , HI , HJ , IJ , JK , IL , KL , AL , and LC

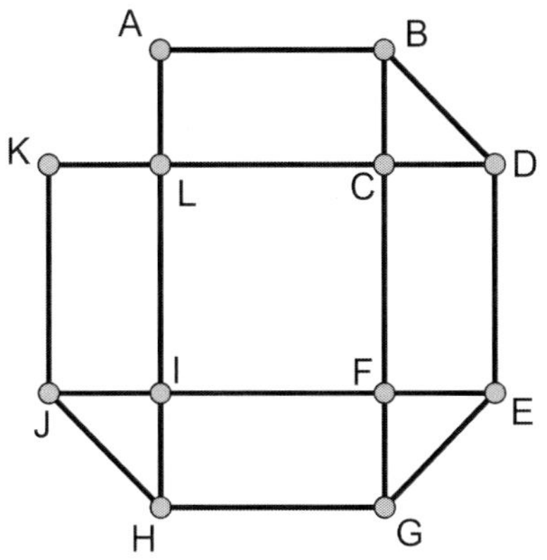

14) Draw an undirected graph with the vertices A,B,C,D,E,F,G,H,I,J,K , and L. Edges as AB , BC , CD , CF , FE , ED , EG , GF , FI , HG , HI , HJ , IJ , JK , IL , KL , AL , and LC

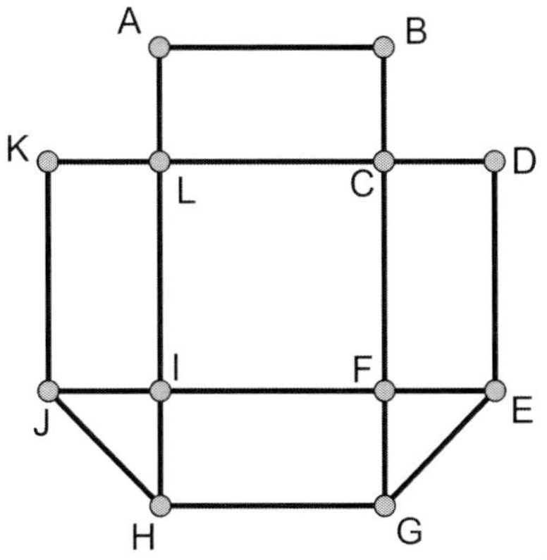

15) Draw an undirected graph with the vertices A,B,C,D,E,F,G,H,I,J,K , and L. Edges as AB , BC , CD , CF , FE , ED , EG , GF , FI , HG , HI , IJ , JK , IL , KL , AL , and LC

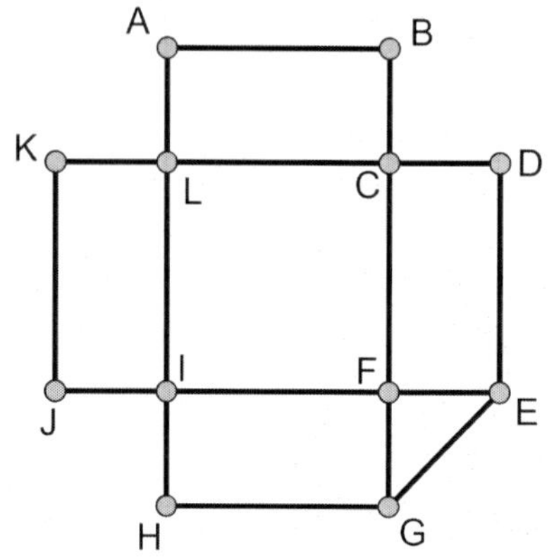

16) Draw an undirected graph with the vertices A,B,C,D,E,F,G,H,I,J,K , and L. Edges as AB , BC , CD , CF , FE , ED , GF , FI , HG , HI , IJ , JK , IL , KL , AL , and LC

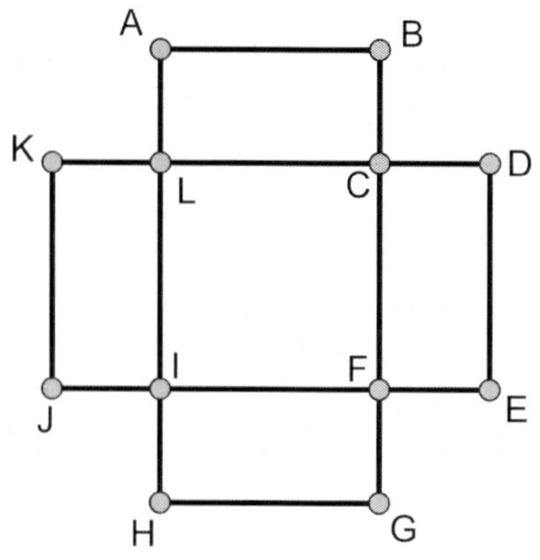

17) Draw an undirected graph with the vertices A,B,C,D,E,F,G,H,I,J,K , and L.
Edges as AB , BC , CD , CF , FE , ED , GF , FI , HG , HI , AF
IJ , JK , IL , KL , AL , and LC

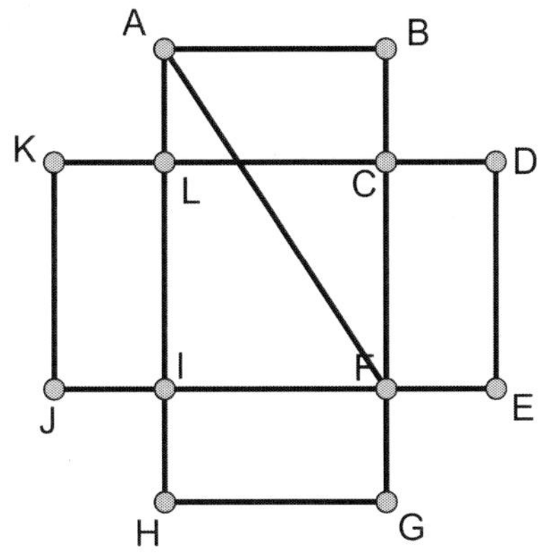

18) Draw an undirected graph with the vertices A,B,C,D,E,F,G,H,I,J,K , and L.
Edges as AB , BC , CD , CF , FE , ED , GF , FI , HG , HI , AF , BI
IJ , JK , IL , KL , AL , and LC

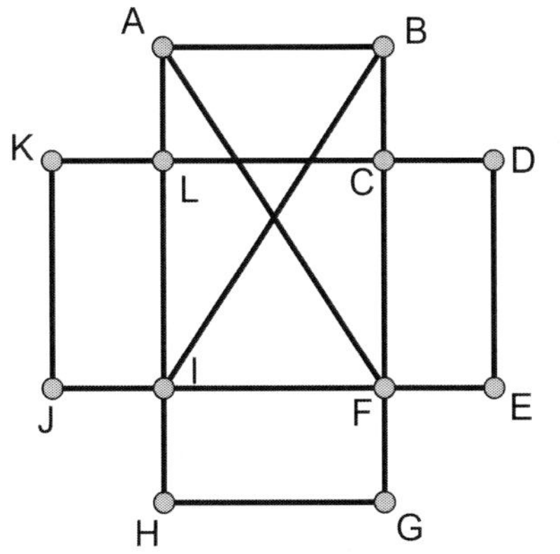

19) Draw an undirected graph with the vertices A,B,C,D,E,F,G,H,I,J,K , and L. Edges as AB , BC , CD , CF , FE , ED , GF , FI , HG , HI , AF , BI , GL IJ , JK , IL , KL , AL , and LC

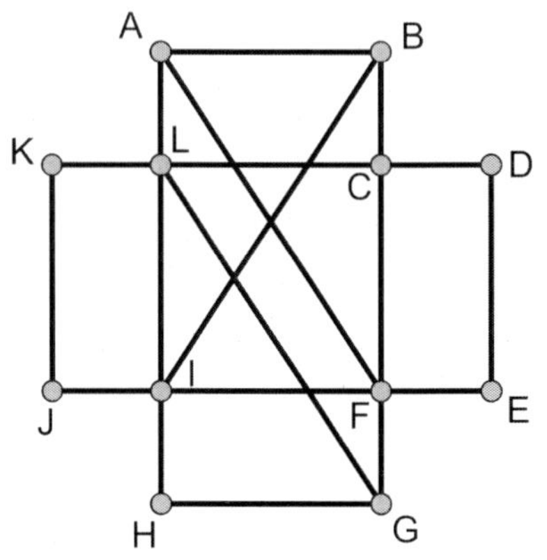

20) Draw an undirected graph with the vertices A,B,C,D,E,F,G,H,I,J,K , and L. Edges as AB , BC , CD , CF , FE , ED , GF , FI , HG , HI , AF , BI , GL , CH IJ , JK , IL , KL , AL , and LC

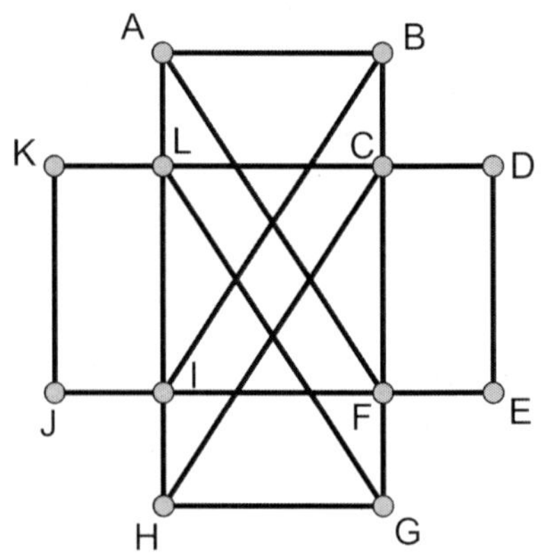

21) How many paths of length equals to 2 exist in the below figure ? List all of them.

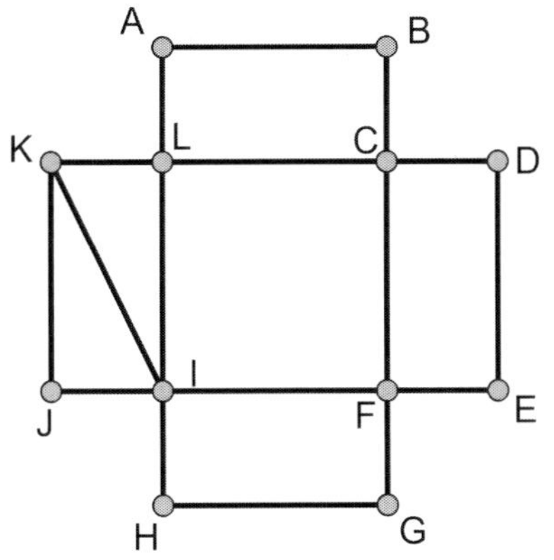

SOLUTION:

ABC , BCD , CDE , DEF , EFG , FGH , GHI , HIJ ,
IJK , JKL , KLA , LAB , LKI , KIJ , KIF , KIH

22) How many paths of length equals to 2 exist in the below figure ? List all of them.

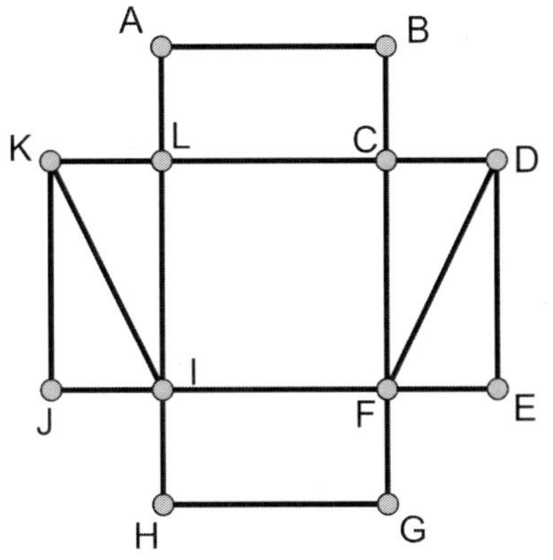

SOLUTION:

ABC , BCD , CDE , DEF , EFG , FGH , GHI , HIJ , CFD , EFD , DFI ,
IJK , JKL , KLA , LAB , LKI , KIJ , KIF , KIH , CDF , EDF , DFG

23) How many paths of length equals to 2 exist in the below figure ?
List all of them.

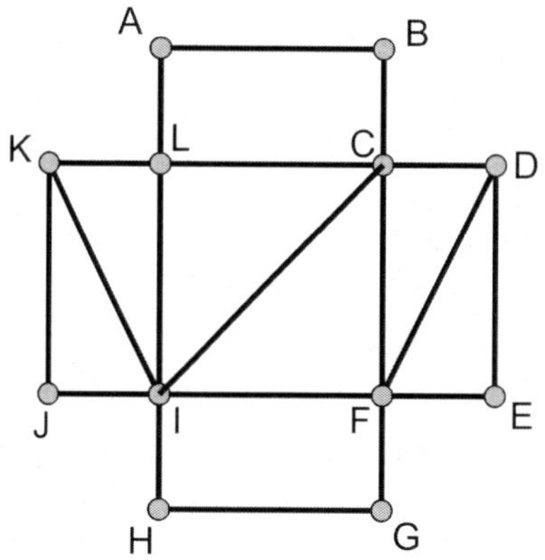

SOLUTION:

ABC , BCD , CDE , DEF , EFG , FGH , GHI , HIJ , CFD , EFD , DFI , IJK , JKL , KLA , LAB , LKI , KIJ , KIF , KIH , CDF , EDF , DFG , LCI , FCI , FIC , LIC , CIH , CIJ , ICB , ICD , CIK

24) How many paths of length equals to 2 exist in the below figure ?
List all of them.

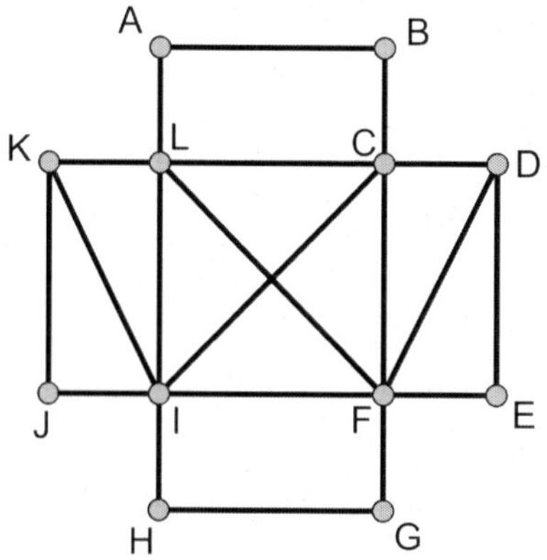

SOLUTION:

ABC , BCD , CDE , DEF , EFG , FGH , GHI , HIJ , CFD , EFD , DFI , IJK , JKL , KLA , LAB , LKI , KIJ , KIF , KIH , CDF , EDF , DFG , LCI , FCI , FIC , LIC , CIH , CIJ , ICB , ICD , CIK , LFE , LFC , LFG , LFI , FLK , FLA , FLC , FLI , LFD

25) How many paths of length equals to 3 originate from the vertex O in the below figure ? List all of them.

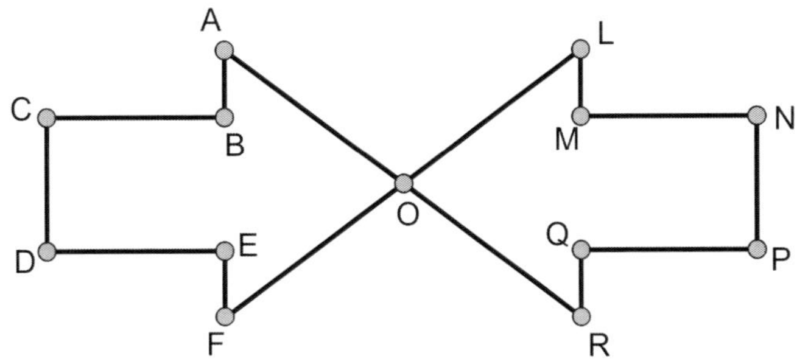

SOLUTION:

ORQP , OLMN , OFED , OABC

26) How many paths of length equals to 3 originate from the vertex O in the below figure ? List all of them.

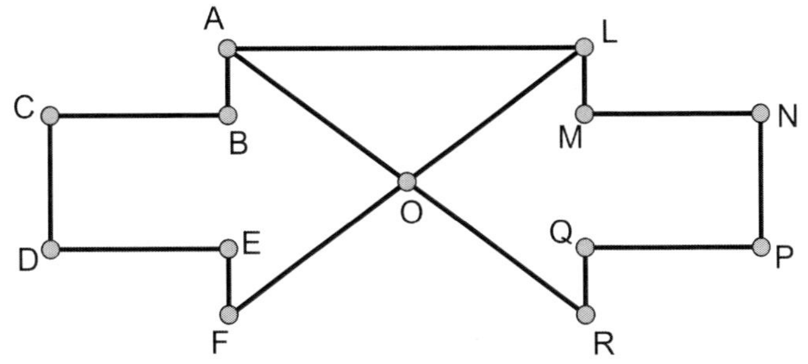

SOLUTION:

ORQP , OLMN , OFED , OABC , OALO , OLAO , OALM , OLMN , OLAB

27) How many paths of length equals to 3 originate from the vertex O in the below figure ? List all of them.

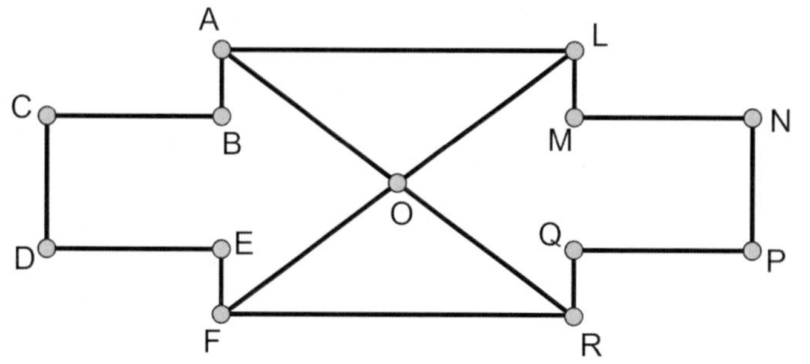

SOLUTION:

ORQP , OLMN , OFED , OABC , OALO , OLAO , OALM , OLMN , OLAB , OFED , OFRQ , OFRO , ORFE , ORFO , ORQP

28) How many paths of length equals to 2 in the below figure ? List all of them.

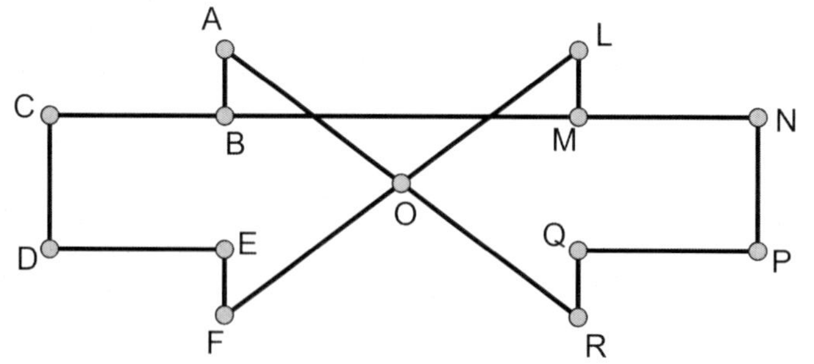

SOLUTION:

ABC , BCD , CDE , DEF , EFO , FOR , ORQ , RQP , QPN , PNM , NMB , NML , MLO , LOF , OAB , OFE , MBC , LMB , ABM

29) How many paths of length equals to 2 in the below figure ? List all of them.

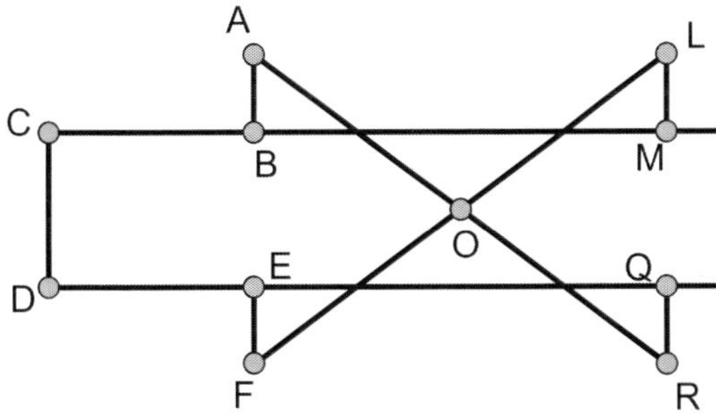

SOLUTION:

ABC , BCD , CDE , DEF , EFO , FOR , ORQ , RQP , QPN , DEQ
PNM , NMB , NML , MLO , LOF , OAB , OFE , MBC , LMB , ABM ,
EQP , FEQ , RQE

30) How many paths of length equals to 2 in the below figure ? List all of them.

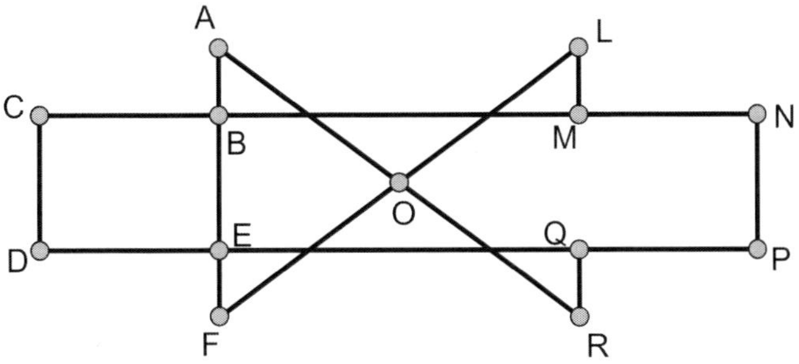

SOLUTION:

ABC , BCD , CDE , DEF , EFO , FOR , ORQ , RQP , QPN , DEQ
PNM , NMB , NML , MLO , LOF , OAB , OFE , MBC , LMB , ABM ,
EQP , FEQ , RQE , ABE , BEF , BEQ , EBM , EBC , BED

31) How many paths of length equals to 2 in the below figure ? List all of them.

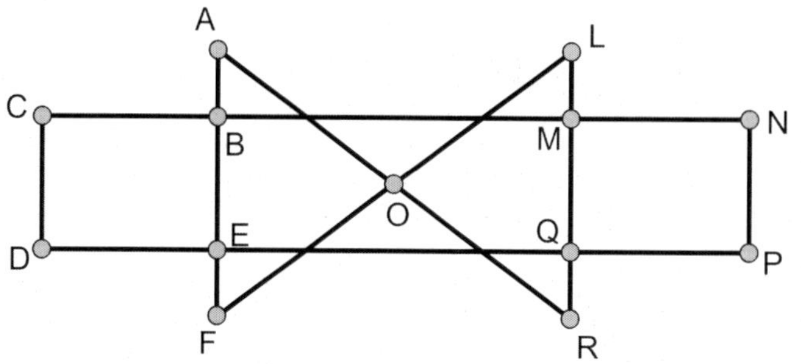

SOLUTION:

ABC , BCD , CDE , DEF , EFO , FOR , ORQ , RQP , QPN , DEQ
PNM , NMB , NML , MLO , LOF , OAB , OFE , MBC , LMB , ABM ,
EQP , FEQ , RQE , ABE , BEF , BEQ , EBM , EBC , BED , LMQ ,
MQR , MQP , MQE , QMB , QML

32) Find if the following graph is traversable or not ?

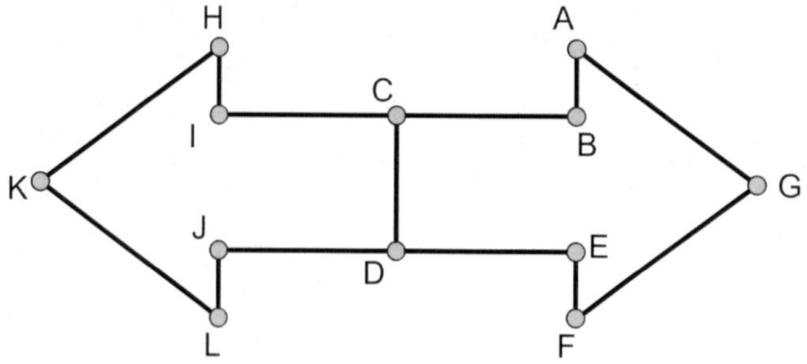

SOLUTION:

The degree of each vertex is given below
A = 2 , B = 2 , C = 3 , D = 3 , E = 2 , F = 2 ,
G = 2 , H = 2 , I = 2 , J = 2 , K = 2 , L = 2
There are 2 odd vertices , so the graph is traversable

33) Find if the following graph is traversable or not ?

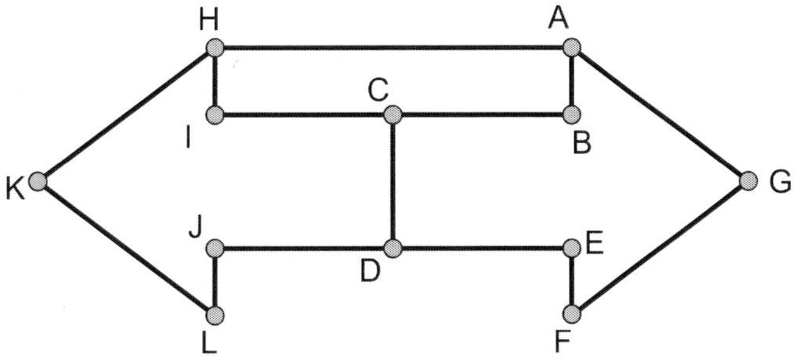

SOLUTION:

The degree of each vertex is given below
A = 3 , B = 2 , C = 3 , D = 3 , E = 2 , F = 2 ,
G = 2 , H = 3 , I = 2 , J = 2 , K = 2 , L = 2
There are 3 odd vertices , so the graph is not traversable

34) Find if the following graph is traversable or not ?

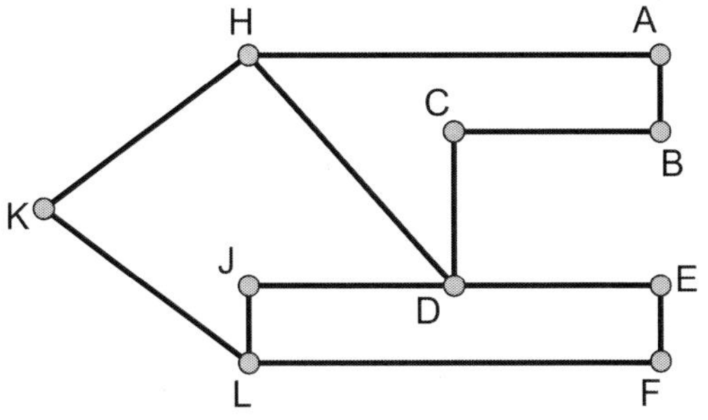

SOLUTION:

The graph is traversable

35) Find if the following graph is traversable or not ?

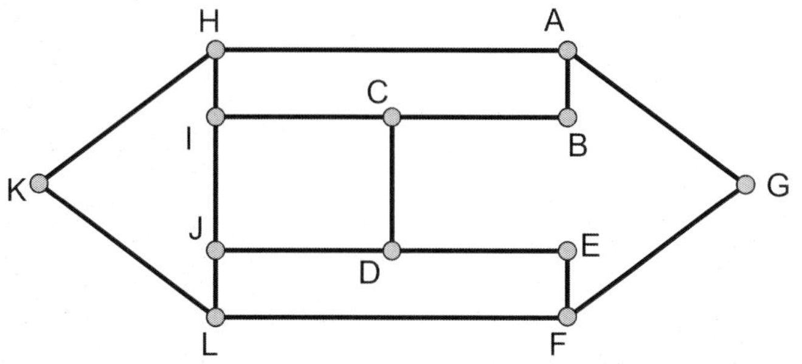

SOLUTION:

The graph is not traversable

36) Find if the following graph is traversable or not ?

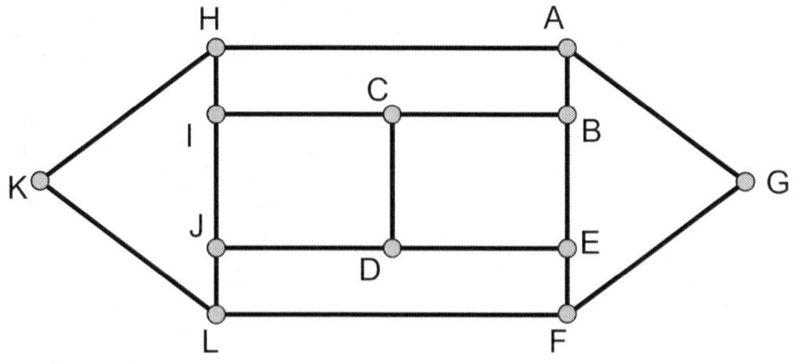

SOLUTION:

The graph is not traversable

37) Find if the following graph is traversable or not ?

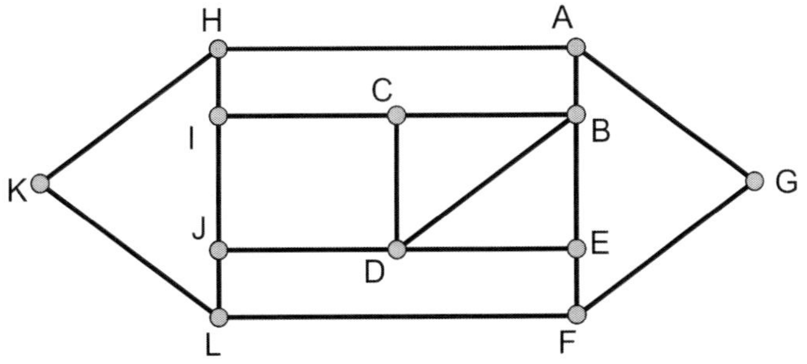

SOLUTION:

The graph is traversable

38) Find if the following graph is traversable or not ?

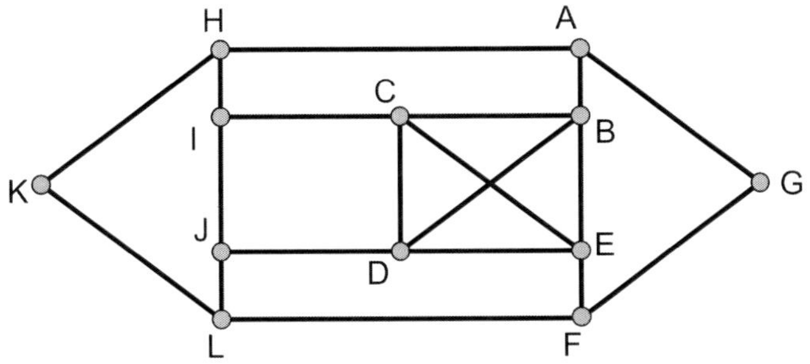

SOLUTION:

The graph is not traversable

39) Find if the following graph is traversable or not ?

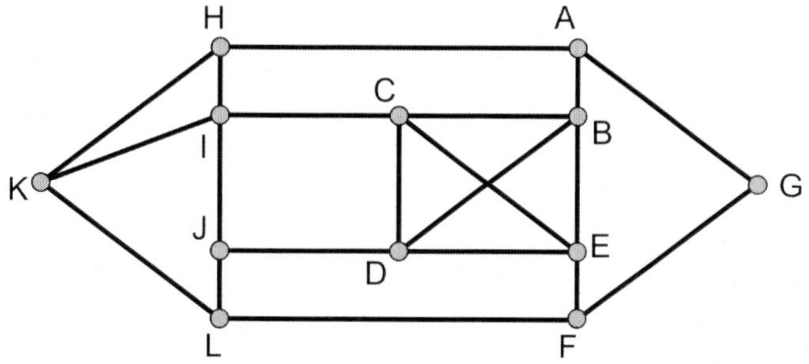

SOLUTION:

The graph is not traversable

40) Find if the following graph is traversable or not ?

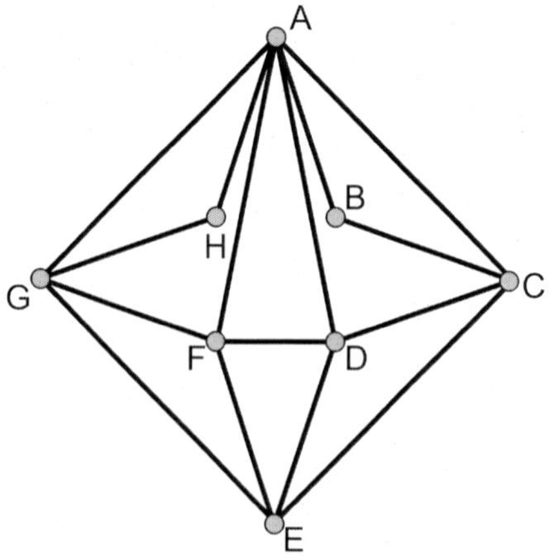

SOLUTION:

The graph is traversable

41) Find if the following graph is traversable or not ?

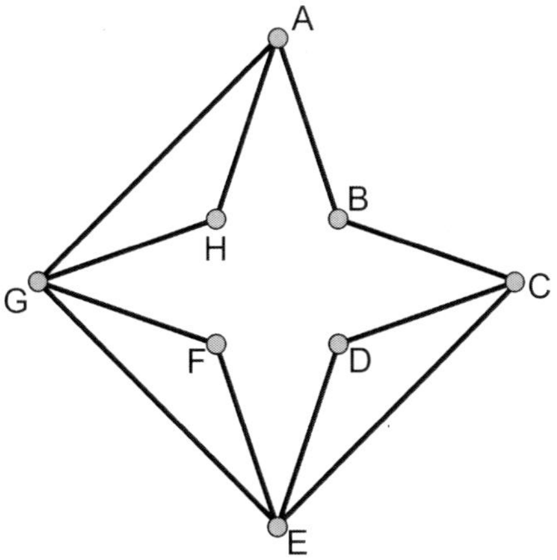

SOLUTION:

The graph is traversable

42) Find if the following graph is traversable or not ?

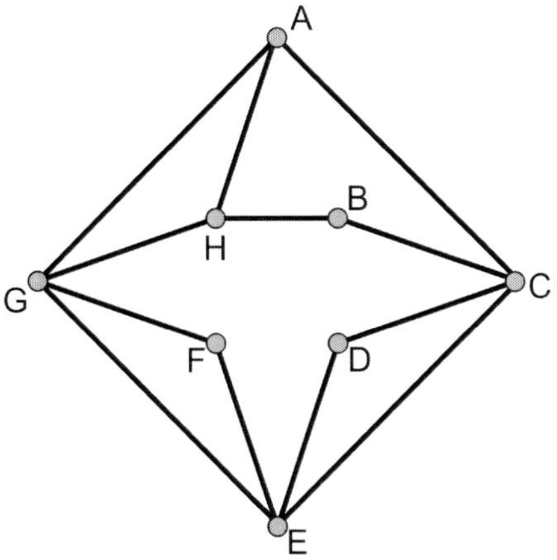

SOLUTION:

The graph is traversable

43) Find if the following graph is traversable or not ?

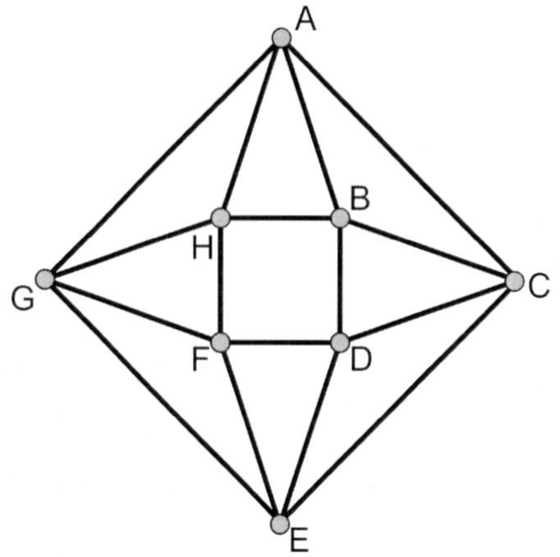

SOLUTION:

The graph is not traversable

44) Find if the following graph is traversable or not ?

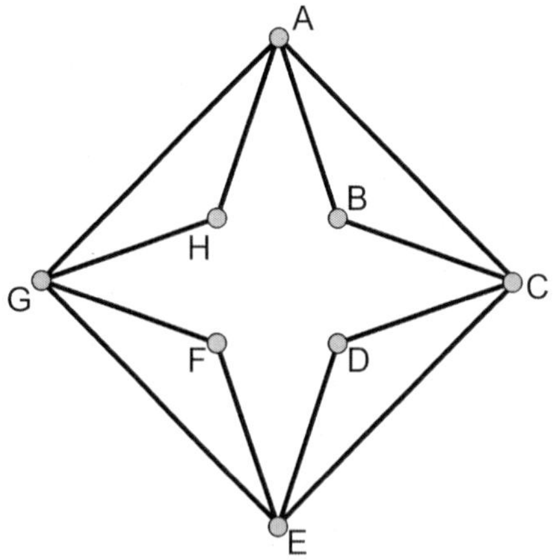

SOLUTION:

The graph is traversable

45) Draw an undirected file with the vertices A,B,C,D,E,F,G,H,I,J,K,L and edges AB , BG , AG , GE , GF , EF , EB , HK , HI , IK , IJ , JK , KL , LJ , HA , IC , CB , JD , DE , LF

SOLUTION:

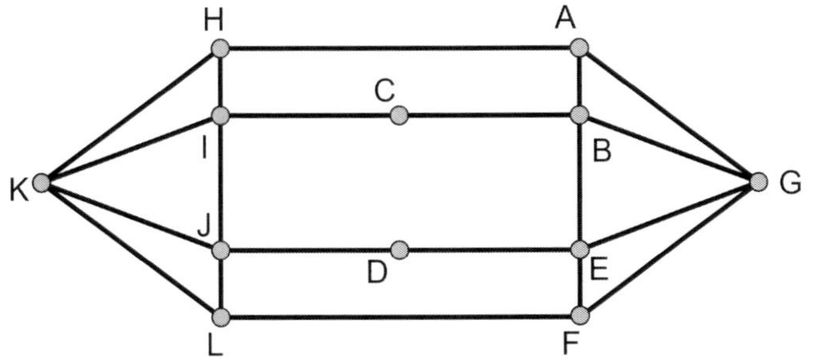

46) Draw an undirected file with the vertices A,B,C,D,E,F,G,H,I,J,K,L and edges AB , BG , AG , GE , GF , EF , EB , HK , HI , IK , IJ , JK , KL , LJ , HA , IC , CB , JD , DE , LF , IE

SOLUTION:

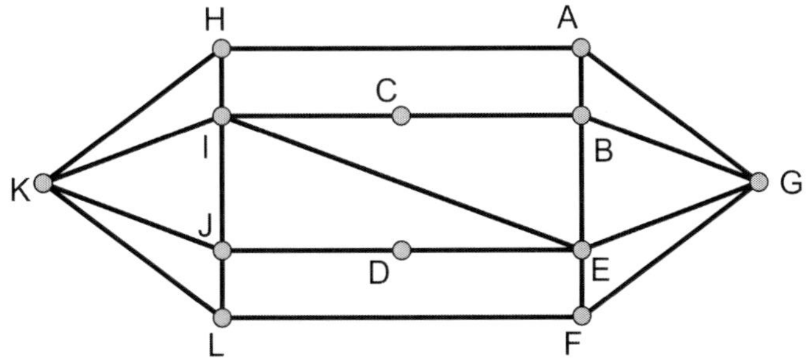

Made in United States
North Haven, CT
05 November 2026

21772733R10107